Endangered Plants

Edited by Sanjeet Kumar

Published in London, United Kingdom

IntechOpen

Supporting open minds since 2005

Endangered Plants
http://dx.doi.org/10.5772/intechopen.87498
Edited by Sanjeet Kumar

Contributors
Nursen Çördük, Cüneyt Akı, Dhatchanamoorthy Narayanasamy, Natesan Balachandran, Muhammad Salihu Abdallah, Rusea Go, Amritpal Singh Kaleka, Gaganpreet Kour Bali, J. Antonio Vázquez-García, Miguel Ángel Muñiz-Castro, Alex Nelson Dahua-Machoa, Edison Osorio-Muñoz, Gerardo Hernandez-Vera, Alóndra Salome Ortega-Peña, Cesar Jacobo-Pereira, Rosa de Lourdes Romo Campos, Noelia Álvarez de Román, Viacheslav Shalisko, Utkarsha Srivastava, Hemant Sood, Jennifer Wilkening, Tiffany J. Pereira, Lara Kobelt, Sameer Quazi, Tanya Golani, Arnaud Martino Capuzzo

First published in London, United Kingdom, 2021 by IntechOpen
IntechOpen is the global imprint of INTECHOPEN LIMITED, registered in England and Wales, registration number: 11086078, 5 Princes Gate Court, London, SW7 2QJ, United Kingdom
Printed in Croatia

British Library Cataloguing-in-Publication Data
A catalogue record for this book is available from the British Library

Additional hard and PDF copies can be obtained from orders@intechopen.com

Endangered Plants
Edited by Sanjeet Kumar
p. cm.
Print ISBN 978-1-83962-893-1
Online ISBN 978-1-83962-894-8
eBook (PDF) ISBN 978-1-83962-899-3

For EU product safety concerns:
IN TECH d.o.o., Prolaz Marije Krucifikse Kozulić 3, 51000 Rijeka, Croatia, info@intechopen.com or visit our website at intechopen.com.

Meet the editor

Dr. Sanjeet Kumar, DELF, is the founder and CEO of Ambika Prasad Research Foundation, India. His current research is focused on status, taxonomy, diversity, phytochemistry, and antimicrobial activity of medicinal and nutraceutical plants; population analysis and restoration of threatened taxa; wetland ecosystems and their bio-wealth; various management practices in protected areas; and training researchers, biologists, and academicians on medicinal plants and threatened taxa. He is chief editor of the Journal of Biodiversity and Conservation. He is documenting the floral and faunal wealth and importance of different landscapes and training undergraduate and postgraduate students in these areas. Dr. Kumar has thirteen years of research experience and published twenty books and 115 research and review articles in journals of national and international repute.

Contents

Preface

Extinction and introduction of new taxa are two crucial biodiversity crises of the modern era. Taxa become endangered and eventually extinct due to the high death rate. There are a number of factors associated with this and many taxa are facing tremendous pressure and are on the verge of extinction throughout the world. Therefore, documentation of endangered species, as well as restoration work in certain landscapes, is very important. *Endangered Plants* presents an account of diversity, threats, restoration, and long-term conservation strategies of endangered plants worldwide. Section I describes endangered species and Section II discusses conservation strategies for these endangered plants.

I express my deep gratitude to the contributors of this book for sharing their valuable works and knowledge on endangered plants. I am also thankful to Mr. Josip Knapić and other staff at IntechOpen for their support in the publication of this book.

I hope this volume will be of use to researchers, scientists, academicians, biologists, and other interested readers.

Sanjeet Kumar
Biodiversity and Conservation Lab., Ambika Prasad Research Foundation,
Bhubaneswar, India

Section 1

Threatened Species

Chapter 1

In Vitro Propagation of *Digitalis trojana* Ivanina., an Endemic Medicinal Plant of Turkey

Nurşen Çördük and Cüneyt Aki

Abstract

Digitalis trojana Ivanina is a member of the Plantaginaceae family and known by its common name, Helen of Troy foxglove. It is perennial endemic to Çanakkale and Balıkesir, northwestern Turkey. In order to develop an efficient shoot regeneration protocol, the leaf explants of *D. trojana* were cultured on Murashige and Skoog (MS) medium containing 6-benzyl adenine (0.1, 0.5, 1.0, 3.0, 5.0 mg/L) and α-naphthalene acetic acid (0.1, 0.5, 1.0 mg/L), 3% (w/v) sucrose and 0.8% (w/v) agar. The highest number of regenerated shoots was obtained from leaf explants that were cultured on MS medium with 3.0 mg/L BA+0.1 mg/L NAA. Regenerated shoots were rooted on MS medium without plant growth regulators. Rooted plants (2–3 cm) were separately transferred to pots containing a mixture of peat and perlite (2:1 v/v) and acclimatized successfully in a growth chamber.

Keywords: endemic, foxglove, *in vitro*, propagation, regeneration

1. Introduction

Turkey has a rich biodiversity as a result of its location, its geological structure and different climatic zones. Turkey hosts three biogeographical regions: Mediterranean, Euro-Siberian and Irano-Turanian. As a result of located on the meeting point of these three different regions, Turkey is one of the most important areas in the world in terms of biological diversity. It is one of the world's richest countries with regard to diversity of plant species, hosting 167 families, 1320 genera and 9996 species [1]. This genetic diversity gains importance especially with the diversity of endemic, rare, medicinal and cultivated plants. A species, or other category of organism that is unique to a defined geographic location such as an island, nation, country, or habitat type is considered endemic to that location. The endemism rate of the Turkish flora is 31.8% and each year new such species are identified. Turkey has 3649 endemic plant species of flora with 31.82% rate of endemism [1–4]. Asteraceae, Lamiaceae, Fabaceae, Caryophyllaceae and Scrophulariaceae families constitute 50% of the distribution of endemic taxa in Turkey [5]. High levels of endemism within Turkey are concentrated in specific areas, such as the Amanos Mountains, the Ilgaz Mountains, the Central Taurus Mountains, the Taşeli Plateau, the Bolkar and Aladağlar Mountains, Kazdagi, Uludağ Mountain, the mountains between Gümüşhane and Erzincan, the Munzur Mountains, and Lake Tuz and its saline steppes [1].

Kazdagi Mountain (Mt. Ida) is in the north western of Turkey, at the transition area of the Euro-Siberian and Mediterranean regions. Kazdagi is the gene center of the west Anatolian region. Endemic and rare taxa have been preserved on the different geological massive and especially in the subalpine zone [6]. The Mountain is ecologically and floristically diverse, containing a number of plant species endemic to Turkey. It has a rich flora thanks to climatic conditions, geological structure and its location. There are about 800 taxa grown naturally in the area and 79 of them are endemic to Turkey. Thirty-one of the endemic taxa are grown only in this Mountain [7–10].

Due to exceptional amount of endemism that brings a huge responsibility to Turkey, it is to ensure that these species are adequately protected from threats or extinction, particularly for those which are related to the crops upon which much of the world depends. Endemic plant species may be endangered due to their distribution specific to a specific region and the decrease in the number of individuals in the population. Endemic plant species should be taken under protection in *ex situ* as well as *in situ* in order for the biodiversity of our country to be passed on to future generations, to prevent the destruction of gene resources, to prevent the loss of wild forms of many cultured species, and for the sustainability of vegetative production. Plant tissue culture is one of the methods used in *ex situ* conservation of plant gene resources. Tissue culture allows the production of new tissue, plant or herbal products by culturing parts of a plant such as cell, tissue, organ, meristem or embryo under aseptic conditions in an appropriate artificial nutrient medium [9, 11, 12]. *In vitro* culture technologies have been increasingly used for *ex situ* conservation of rare or endangered endemic plants [13, 14]. Besides, high amounts of production of medicinal plants and their metabolites can be achieved via *in vitro* culture [15–17].

The genus *Digitalis* L., commonly known as the foxglove, is a member of the Plantaginaceae and encompasses approximately 20 species. *Digitalis* species are biennial or perennial herbs, rarely small shrubs with simple, alternate leaves, which are often crowded in basal rosettes [18, 19]. This genus is native to western and southwestern Europe, western and central Asia, Australasia and northwestern Africa. *Digitalis* species is represented by eight species and two subspecies, including four endemic, in Turkey. Among these species, *Digitalis trojana* Ivanina is known by its common name, Helen of Troy foxglove. It is perennial endemic to Kazdagi, northwestern Turkey [20]. According to the Red Data Book of Turkish Plants, conservation status of this species is declared as vulnerable (VU) [21]. Digitalis species are a medicinally important group because of the cardenolides or cardiac glycosides (CGs) applied in human medicine. The cardenolides have been used as the most effective heart drugs for treating several heart defects [22–24]. They also show a broad spectrum of biological activities, containing anti-carcinogenic, acaricidal and antibacterial properties. Recent studies are especially focused on the anticarcinogenic effects of digoxin and digitoxin particularly for cancer treatments [25, 26]. There are lots of reports about *in vitro* cultures of *Digitalis* species [22, 27–31]. The aim of this research is to carry out an efficient *in vitro* propagation of *D. trojana* for *ex situ* conservation.

2. Materials and methods

2.1 Plant materials

Plant seeds of *D. trojana* were collected from the Kazdagi National Park, Turkey (**Figure 1**). Plant samples were prepared as herbarium materials and voucher

Figure 1.
Plant samples of Digitalis trojana (a, b, c).

specimens were deposited in the Herbarium of Çanakkale West Anatolia (CBB, Çanakkale, Turkey) under number 351, 352 and 353.

Seeds were surface-sterilized by stirring in 3% (v/v) solution of sodium hypochlorite containing two drops of 0.1% Tween 20 for 20 min. Under sterile conditions, followed by 5 times rinsed in sterile water. The surface-sterilized seeds were cultured on MS basal medium (MS: M0222, Duchefa Biochemie B.V., Haarlem, Netherlands) [32], containing 3% (w/v) sucrose and 0.8% (w/v) agar, and keep in the dark for a week for successful germination. Germinated seedlings were subsequently transferred to a growth chamber at 25 ± 2°C under a 16-h light and 8-h dark long day cycle with a light intensity of 72 μmol m^{-2} s^{-1}provided by cool-white fluorescent lamps.

2.2 Induction of adventitious shoots

Shoot cultures of *D. trojana* were established *in vitro* using leaf segments (5x5 mm) excised aseptically from 12-week-old *in vitro* germinated seedlings. The explants were cultured on MS medium supplemented with BA (0.1, 0.5, 1.0, 3.0, 5.0 mg/L) in combination with NAA (0.1, 0.5, 1.0 mg/L) and 3% (w/v) sucrose (S0809, Duchefa). All media were gelled with 0.8% (w/v) agar (P1001, Duchefa) and the pH was adjusted to 5.75 before autoclaving. All cultures were maintained at 25 ± 2°C under the 16/8 h photoperiod with a light intensity of 72 μmol m^{-2} s^{-1}. Ten explants were cultured per petri dish for each type of explant, and at least five replicates were used for each treatment. The mean number of regenerated shoots per explant was recorded in each culture after twelve weeks.

2.3 Shoot multiplication, rooting and transfer of plantlets to soil

In order to multiply the shoots, the well-developed shoots (ca. 1 cm in length) grown for 8 weeks on shoot induction medium were excised and transferred to fresh MS medium containing the same PGRs as the shoot induction medium. For rooting, the shoots (2–3 cm in length) produced after multiplication were separately transferred to basal MS medium, 3% (w/v) sucrose and 0.8% (w/v) agar without plant growth regulators. All the plants with well-developed roots were taken out of the solid medium, gently washed in running water and shortly after transplanted into pots (5 mm in diameter) containing a mixture of peat and perlite (2:1 v/v). Plantlets were covered with plastic cups to maintain a high humidity and a few holes were opened on the cups. The diameter of the holes was gradually increased over the next 2 weeks. After one month, the plantlets were uncovered and were then transplanted to new pots (15 mm in diameter) containing the mixture of 2 peat:1 perlite.

2.4 Statistical analysis

The mean number of shoots and percent of explants forming shoots were analyzed after twelve weeks from initial inoculation for all explants. All data were evaluated by analysis of variance and mean values were compared using MINTAB. The interaction of plant growth regulators was analysis with MSTAT. The statistic model ($Yijk = \mu + Ai + Bj + AB\,ij + \Sigma\,ijk$) was used for assigning the effects of plant growth regulator's concentration on shoot regeneration.

3. Results and discussion

The explants cultured on MS medium without plant growth regulators were only slightly expanded callus and shoot formation wasn't observed in this medium. Adventitious shoots were formed out of the cut edges of the explants, cultured on MS medium containing BA in combination with NAA within three weeks in the culture (**Figure 2**). Mean number of shoots per explants and percent of explants forming shoots were recorded (**Table 1**). The highest shoot formation was obtained on media containing 3.0 mg/L BA+0.1 mg/L NAA and 5.0 mg/L BA+1.0 mg/L NAA. Although the number of regenerated shoots are higher on media containing 5.0 mg/L BA+1.0 mg/L NAA, the vitrification was occurred in the shoots in this medium and the shoots were not multiplicated. As a result of analysis of variance, it was shown that the effect of NAA varied depending on BA concentration for shoot regeneration at 12th week of culture ($P < 0.01$).

BA was promoted to shoot formation by interaction with NAA (**Figure 3**). The ratio of NAA to BA was also very significant. Especially, 3.0 or 5.0 mg/L concentration of BA was effective for shoot formation from *D. trojana* leaf explants. Our results are consistent with previous report on *D. thapsi* L. [30], which reported that the presence of high concentrations of BA (3, 4, 5 mg/L) in combination with IAA or 2,4 D or NAA promoted callus formation and shoot organogenesis from leaf explants. In another research, BA promoted adventitious bud differentiation alone, but addition of auxin significantly increased the bud forming capacity of leaf explants of *D. minor* L. [33]. The best shoot proliferation was observed among

Figure 2.
Morphogenic callus induction and shoot regeneration in D. trojana from leaf explant culture (a), multiple adventitious shoot regeneration on MS medium with 3.0 mg/L BA +0.1 mg/L NAA (b, c, d).

Plant growth regulators (mg/L)		Percent of explants forming shoots	Mean number of shoots per explants
BA	NAA		
—	—	0.0	0.0
0.1	0.1	4.0	0.3 ± 0.1 ^{Da}
0.5	0.1	16.0 ± 0.3	2.3 ± 0.5 ^{CDa}
1.0	0.1	20.0 ± 1.4	10.3 ± 0.3^{BCa}
3.0	0.1	32.0 ± 0.1	28.0 ± 1.8^{Aa}
5.0	0.1	45.0 ± 0.5	16.3 ± 0.1 ^{Bb}
0.1	0.5	0.0	0.0 ^{Ba}
0.5	0.5	0.0	0.0 ^{Ba}
1.0	0.5	0.0	0.0 ^{Bb}
3.0	0.5	32.0 ± 0.9	17.3 ± 0.3^{Ab}
5.0	0.5	12.0 ± 0.7	2.3 ± 0.6^{Bc}
0.1	1.0	0.0	0.0 ^{Ba}
0.5	1.0	13.3 ± 1.6	0.6 ± 0.1^{Ba}
1.0	1.0	12.0 ± 1.8	1.6 ± 0.5^{Bb}
3.0	1.0	4.0 ± 0.2	0.3 ± 0.1^{Bc}
5.0	1.0	14.6 ± 0.5	26.3 ± 3.0^{Aa}

Variety of BA concentration is significant in every NAA concentration and variety of NAA concentration is significant in every BA concentration (means with the same letters *are not significantly different at P < 0.01).*

Table 1.
Shoot regeneration from leaf explants cultured on MS medium containing different concentration of BA and NAA.

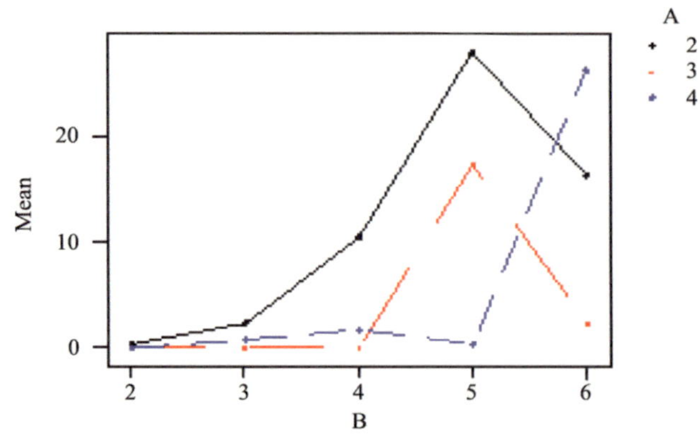

Figure 3.
Interaction of BA and NAA (A2; 0.1 mg/L NAA, A3; 0.5 mg/L NAA, A4; 1.0 mg/L NAA; B2; 0.1 mg/L BA; B3; 0.5 mg/L BA; B4; 1.0 mg/L BA; B5; 3.0 mg/L BAP; B6; 5.0 mg/L BA) (Y axis: The mean number of shoots).

explants cultured on MS medium with 3.0 mg/L BA +0.1 mg/L NAA (**Figure 4**). These 2–3 cm long elongated shoots were transferred to root formation medium, MS without plant growth regulators within 1 week. MS medium containing 0.5 mg/L NAA was also used as root forming medium but on this medium roots were not induced.

Figure 4.
Multiple adventitious shoots on MS medium with 3.0 mg/L BA + 0.1 mg/L NAA (a-d).

At the end of experimental series of our research, plantlets were adapted to *ex vitro* conditions and then transplanted to plastic flowerpots (**Figure 5**). Finally, plantlets were successfully acclimatized in the greenhouse after they were taken to the plastic cups containing 2 peat:1 perlite (**Figure 6**). All *in vitro* regenerated plantlets were grown healthy.

Figure 5.
Regenerated shoots in elongation medium (a), shoots developed their roots (b), regenerated plantlets (c), plantlets with well-developed roots transferred for acclimatization to pots (d).

Figure 6.
Plantlets taken to the plastic cups containing 2 peat:1 perlite (a-d).

4. Conclusions

In conclusion, we attained an effective, rapid and high-efficient *in vitro* propagation protocol for mass production of *Digitalis trojana* plants. We showed that leaf explants can be cultured on MS medium with 3.0 mg/L BA+0.1 mg/L NAA with a satisfactory frequency of plant regeneration of *Digitalis trojana*. The results of this study demonstrate that this protocol is a reliable method for future studies on this endemic species.

Acknowledgements

This work was a part of the Nurşen Çördük's PhD thesis and supported by Scientific Research Projects Commission of Çanakkale Onsekiz Mart University (2008/26).

Author details

Nurşen Çördük and Cüneyt Aki*
Department of Biology, Faculty of Sciences and Arts, Çanakkale Onsekiz Mart University, Çanakkale, Turkey

*Address all correspondence to: cuneytaki@gmail.com

References

[1] FAO. Biodiversity of Turkey. Contribution of genetic resources to sustainable agriculture and food systems: Ankara; 2018. 222 p. Licence: CC BY-NC-SA 3.0 IGO.

[2] Avcı M. Çeşitlilik ve endemizm açısından Türkiye'nin bitki örtüsü. İstanbul Üniversitesi Edebiyat Fakültesi Coğrafya Dergisi. 2005;**13**:27-55

[3] Güner A, Aslan S, Ekim T, Vural M, Babaç MT, editors. Türkiye bitkileri listesi (Damarlı Bitkiler), Nezahat Gökyiğit Botanik Bahçesi ve. Flora Araştırma Derneği Yayını: İstanbul; 2012 1290 p

[4] Demirayak F. Biyolojik çeşitlilik-doğa koruma ve sürdürülebilir kalkınma. TÜBİTAK, Vizyon 2023, Biyolojik Çeşitliliğin Korunması ve Sürdürülebilir Kalkınma, Aralık 2002 [Internet]. Available from: http://www.tubitak.gov.tr /tubit ak_content_files /vizyon2023/csk/EK-14.pdf.

[5] Şenkul Ç, Kaya S. Türkiye endemik bitkilerinin coğrafi dağılışı. Türk Coğrafya Dergisi. 2017;**69**:109-120

[6] Uysal I. An overview of plant diversity of Kazdagi (Mt. Ida) Forest National Park. Turkey. Journal of Environmental Biology. 2010;**31**:141-147

[7] Davis PH. Flora of Turkey and the east aegean islands. Vol.6. Edinburg: Edinburg University Press; 1972. pp. 680-687

[8] Dirmenci T, Satıl F, Tümen G. Kazdağı Milli Parkı çiçekli bitkileri. Balıkesir: Zeytinli Belediyesi; 2007. pp. 182-185

[9] Gürel E, Türker AU. Organogenesis. In: Babaoğlu M, Gürel E, Özcan S, editors. Bitki Biyoteknolojisi: Doku Kültürü ve Uygulamaları. Selçuk Üniversitesi Vakfı Yayınları: Konya. 2001. pp. 36-70

[10] Tümen G, Satıl F, Duman H, Başer KHC. Two new Records for Turkey: *Satureja icarica* P. H. Davis, *S. pilosa* Velen. Turkish Journal of Botany. 2000;**24**:211-214

[11] Babaoğlu M, Yorgancılar M, Akbulak AM. Doku Kültürü: Temel laboratuar teknikleri. In: Biyoteknolojisi B, editor. Babaoğlu, M., Gürel, E. ve Özcan, S. Doku Kültürü ve Uygulamaları, Selçuk Üniversitesi Vakfı Yayınları: Konya; 2001. pp. 1-35

[12] Trigiano RN, concept GDJP t c, exercises l. CRS Pres. Boca Raton: Florida. 1996:472

[13] Sasaran V, Cripps R, Ramsay MM, Atherton C, McMichien M, Prendergast G, et al. Conservation *in vitro* of threatened plants-progress in the past decade. In Vitro Cellular & Developmental Biology - Plant. 2006;**42**:206-214

[14] George EF, Hall MA, Klerk GD. Plant propagation by tissue culture. 3rd. ed. Springer; 2007. p. 30-43.

[15] Constabel F. Medicinal plant biotechnology. Planta Medica. 1990;**56**:421-425

[16] Rout GR, Samantaray S, Das P. *In vitro* manipulation and propagation of medicinal plants. Biotechnol. Adv. Biotechnology Advances. 2000;**18**:91-120

[17] Chaturvedi HC, Jain M, Kidwai NR. Cloning of medicinal plants through tissue culture-a review. Indian Journal of Experimental Biology. 2007;**45**(11):37-48

[18] Bräuchler C, Meinberg H, Heubl G. Molecular phylogeny of the genera *Digitalis* L. and *Isoplexis* (Lindley) Loudon (Veronicaceae) based on ITS- and trnL-F sequences. Plant Systematics and Evolution. 2004;**248**:111-128

[19] Albach DC, Meudt HM, Oxelman B. Piecing together the "new" Plantaginaceae. American Journal of Botany. 2005;**92**:297-315

[20] Uysal İ, Ozturk M. *Digitalis trojana* Ivan. Endemik türünün morfolojisi, anatomisi ve ekolojisi. Anadolu University Journal of Faculty of Arts and Sciences. 1991;**3**:53-61

[21] Ekim T, Koyuncu M, Vural M, Duman H, Aytac Z, Adıguzel N. Red Data Book of Turkish Plants (Pteridophyta and Spermatophyta). Ankara: Foundation for Turkish Nature Conservation and Van Centinential University Press; 2000 246 p

[22] Herrera MT, Cacho M, Corchete P, Fernandez-Tarrago J. One step shoot multiplication and rooting of *Digitalis thapsi* L. Plant Cell, Tissue Organ Culture. 1990;**22**:179-182

[23] Haux J. Digitoxin is a potential anticancer agent for several types of cancer. Medical Hypotheses. 1999;**53**(6):543-548

[24] Winnicka W, Bielawski K, Bielawska A. Cardiac glycosides in cancer research and cancer therapy. Acta Poloniae Pharmaceutica. 2006;**63**:109-115

[25] Padua R, Waibel R, Kuate SP, Schebitz PK, Hahn S, Gmeiner P, et al. A simple chemical method for synthesizing malonyl hemiesters of 21-hydroxypregnanes, potential intermediates in cardenolide biosynthesis. Steroids. 2008;**73**:458-465

[26] Elbaz H, Stueckle T, Tse W, Rojanasakul Y, Dinu C. Digitoxin and its analogs as novel cancer therapeutics. Experimental Hematology & Oncology. 2012;**1**(1):4

[27] Erdei I, Kiss Z, Maliga P. Rapid clonal multiplication of *Digitalis lanata* in tissue culture. Plant Cell Reports. 1981;**1**:34-35

[28] Perez-Bermudez P, Brisa MC, Cornejo MJ, Segura J. *In vitro* morphogenesis from excised leaf explants of *Digitalis obscura* L. Plant Cell Reports. 1984;**3**:8-9

[29] Matsimoto M, Koga S, Shoyama Y, Nıshıoka I. Phenolic glycoside composition of leaves and callus cultures of *Digitalis purpurea*. Phytochemistry. 1987;**26**(12): 3225-3227

[30] Vela S, Gavidia I, Perez-Bermıjdez P, Segura J. Micropropagation of juvenile and adult *Digitalis obscura* and cardenolide content of clonally propagated plants. In Vitro Cellular & Developmental Biology - Plant. 1991;**27**:143-146

[31] Sales E, Nebauer SG, Arrillaga I, Segura J. Plant hormones and *Agrobacterium tumefaciens* strain 82.139 induce efficient plant regeneration in the cardenolide-producing plant *Digitalis minor*. Journal of Plant Physiology. 2002;**159**:9-16

[32] Murashige T. Skoog F. a revised medium for rapid growth and bioassay with tobacco tissue culture. Physiologia Plantarum. 1962;**15**:473-497

[33] Cacho M, Moran M, Herrera MT, Fernandez-Tarrago J, Corchete MP. Morphogenesis in leaf, hypocotyls and root explants of *Digitalis thapsi* L. cultured *in vitro*. Plant Cell, Tissue and Organ Culture. 1991;**25**:117-123

Chapter 2

How to Save Endangered Magnolias? From Population Biology to Conservation Action: The Case of Allopatric Radiation in Western Mexico

José Antonio Vázquez-García, Miguel Angel Muñiz-Castro, Alex Dahua-Machoa, Edison Antonio Osorio-Muñoz, Gerardo Hernández-Vera, Alondra Salomé Ortega-Peña, Rosa de Lourdes Romo-Campos, César Jacobo-Pereira, Noelia Álvarez de Román and Viacheslav Shalisko

Abstract

Numerous Neotropical *Magnolia* species are endangered and red-listed by the IUCN. Here we highlight major results from over a decade of research on endangered magnolias in western Mexico. Particularly, we compare three species of *Magnolia* (*M. pugana*, *M. pacifica* and *M. vallartensis*) along a large-scale continentality and moisture gradients, in terms of a) their morphological adaptations, b) genetic structure, diversity, and differentiation, c) reproductive phenology, and d) floral scents and their floral visitors. Fieldwork along this gradient unveiled two new species of *Magnolia* sect. *Magnolia; M. granbarrancae* and *M. talpana*. We found that most continental populations have a higher extinction risk than those with greater maritime influence, due to their lower genetic diversity, and greater fragmentation, isolation, and water stress. Also, these populations are more vulnerable to the environmental conditions predicted with the global warming climate scenarios. We share fieldwork experience and advise on pre-germination treatments and seed dormancy. We propose an ex-situ and in-situ conservation strategy, identify new challenges, and suggest future directions of collaborative work as a global *Magnolia* conservation consortium.

Keywords: allopatric radiation, conservation, floral scents, genetic diversity and differentiation, new species, species distribution modeling, pre-germination treatments

1. Introduction

Magnolias have captivated mankind since ancient times, their lush evergreen foliage, varied floral scents and spectacular flowers make them highly attractive.

Distributed from temperate to tropical areas of the Americas and Asia, they have an extraordinary value in horticulture, they are harvested from natural forests, either as a beauty item for solemn ceremonial rites or as a source of wood for construction, food, and traditional medicine [1, 2].

The family Magnoliaceae Juss. [3] belongs to one of the basal clades in the angiosperms, it consists of 350 species, two genera, *Liriodendron* L. [4] and *Magnolia* L. [4] and 15 sections [5]. It is disjunct and bicontinental; in the America and SE Asia, from near sea level to 3400 m [6].

The fossil record suggests a North American origin for the family, migrating east through the Disko Island, Greenland, and Thulean Landbridge, with a range expansion into Europe and eventually to Asia. The cooling of the climate during the mid-Cenozoic resulted in a migration to lower latitudes and the extinction of Magnoliaceae in Europe and Siberia. Finally, the Ice-house conditions of the late-Cenozoic drove the group further south to highlands of Central and South America, as well as the mountains in South-east Asia [7].

In Mexico, as a result of allopatric speciation, there are 36 species in three sections of *Magnolia*: *M.* section *Macrophyllae* Figlar & Noot. [8] with five species of dorsal dehiscence [1, 9], *M.* sect. *Talauma* (Juss.) Baill. [3, 10] with 13 species mostly of ventral dehiscence [2] and *M.* sect. *Magnolia* with 18 species of dorsal dehiscence [11, 12].

In western Mexico, there are eight species of *Magnolia* [13], two belong to *M.* sect. *Talauma*: *M. jaliscana* A. Vázquez & R. Guzmán [14] and *M. ofeliae* A. Vázquez & Cuevas [15]— both confined to the lower slopes of the Pacific Lowlands province at the edge of the Sierra Madre del Sur province— and six belong to *M.* sect. *Magnolia*: *M. iltisiana* A. Vázquez [11], from the highlands of the Sierra Madre del Sur; *M. pacifica* A. Vázquez [11] from the western end of the Sierra Madre del Sur province to the western end of the Transmexican Volcanic Belt province; *M. pugana* (Iltis & A. Vázquez) A. Vázquez & Carvajal [16] from the canyons and valleys north of Guadalajara city, in the Transmexican Volcanic Belt; *M. vallartensis* A. Vázquez & Muñiz-Castro [14] in the Pacific Lowland province southeast of Puerto Vallarta; and the two species here described as new (**Figure 1**).

Here, we mostly focus in the species which constitute the *Magnolia pacifica* species complex [11], consisting of species with glabrous ellipsoid fruits distributed from tropical climate lowlands to temperate climate mountains in the confluence of four biogeographic provinces Sierra Madre Occidental, Sierra Madre del Sur, Pacific Lowlands and Transmexican Volcanic Belt [17] (**Figure 1**). Populations of the Barrancae Group, occurring in the Río Verde region in the Zapotlanejo e Ixtlahuacán del Río municipalities (north of Guadalajara city) [18] are here described as a new species (*M. granbarrancae*). Similarly, populations from the mountains of southern Talpa de Allende in W Jalisco are here described as a new species: *M. talpana;* while several populations from northern Nayarit, considered morphologically distinct from *M. pacifica* sensu stricto are here designated as the Huajicori group.

Magnolias in western Mexico, occur in four of its six biogeographic provinces, isolated in canyons ("barrancas") or protected ravines of mountains, with minimum if any gene flow among their populations. They have undergone allopatric radiation, diverging in morphology, genetics, phenology, floral scents, and occupy different ecological niches (**Figure 1**) [11, 14, 18–20].

Here we aim to compare three species of *Magnolia* along a large-scale continentally and moisture gradients, in terms of (a) their morphological adaptations, (b) genetic structure, diversity, and differentiation, (c) reproductive phenology, and (d) floral scents and visitors; we also present an extinction risk assessment of each of the species and their vulnerability to predicted global warming climate scenarios.

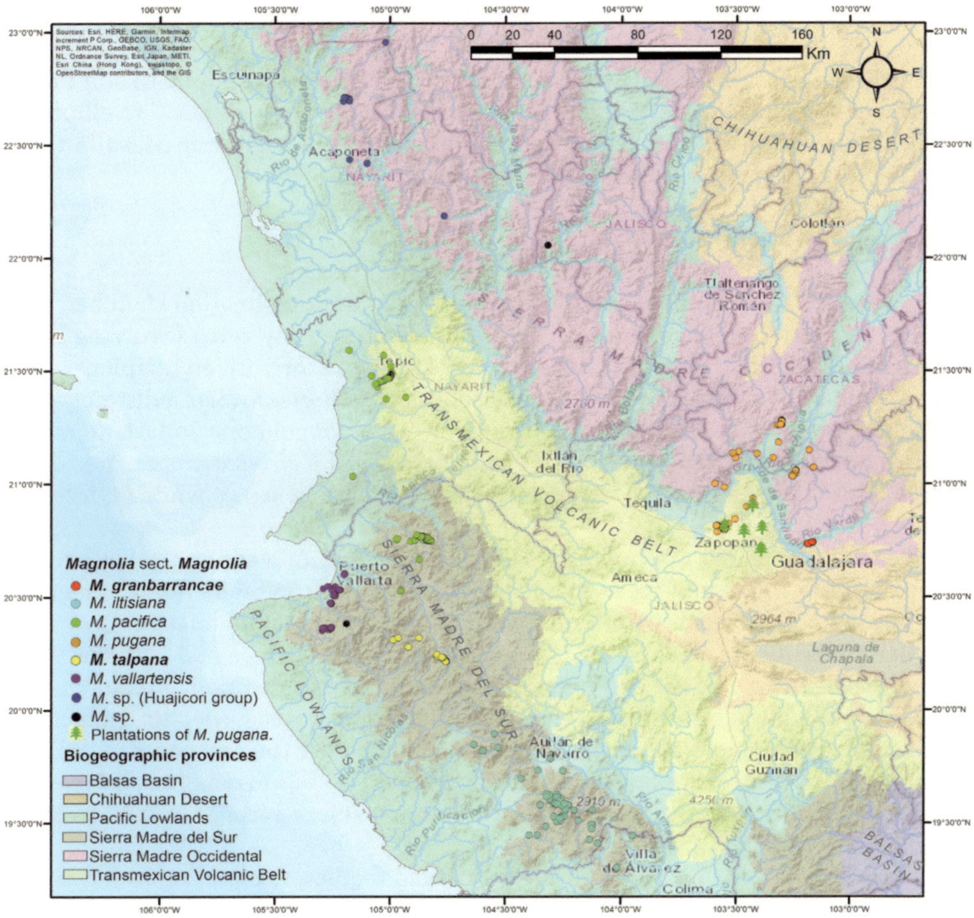

Figure 1.
Current distribution of Magnolia *sect.* Magnolia *in western México and their occurrence in biogeographic provinces [17]. Except for* M. iltisiana *all belong to the* Magnolia pacifica *complex. Undetermined* Magnolia (M. *sp).*

2. Population biology

2.1 Geographic distribution

Seven morphological distinct taxa of the *Magnolia pacifica* species complex occur in western Mexico (**Figure 1**), however most comparisons in this study include the three species growing along a continentality and moisture gradient, from west to east: *M. vallartensis*, *M. pacifica* and *M. pugana* and only few comparisons (morphological and genetical) also include discussion of the *M. granbarrancae* and *M. talpana* here described as new.

2.2 Morphological differentiation and adaptations

2.2.1 Habit and foliage

All species in the *Magnolia pacifica* complex are evergreen long leaved and usually sympodial and tortuous trees. *M. vallartensis* and the new *M. granbarrancae* are generally small trees (8.0–15.0(−23.0) m) with smaller trunk diameter at breast height (1.3 m) (0.2–0.5 m), while *M. pugana*, *M. pacifica* and the new *M. talpana* are large trees (15.0–35.0 m) with thicker trunks, (0.3–1.6 m) in diameter at breast height.

The leaves in *Magnolia vallartensis* are distinctively broadly elliptic to elliptic or occasionally oblong, while the other four species are narrowly elliptical to elliptical to oblanceolate. The width to length ratio of mature leaves is larger in *M. vallartensis* (45–51%) than that of the other species *M. pacifica* (30–39%).

2.2.2 Flowers and floral visitors

Flowers of *Magnolias* in western Mexico differ in their pollination chamber, whether having a fully distinct chamber at female phase (day zero) involving all petals, like in *M. vallartensis*, *M. talpana* and *M. iltisiana,* or with an incipient or loose chamber involving mostly the inner petals. *M. vallartensis*, has a distinct subglobose pollination chamber, *M. iltisiana* has an oblongoid one and *M. talpana* has the narrowest oblongoid and tight chamber. *Magnolia granbarrancae* and *M. pugana* have the greater width to length ratio on petals and sepals while *M. talpana* has the smallest one.

Like other basal angiosperm lineages such as Araceae, Arecaceae, Cyclanthaceae, Nymphaeaceae and Annonaceae, the family Magnoliaceae exhibits floral traits that have been hypothesized as evolutionary adaptations to beetle pollination (cantharophily) [21]. In general, large, bisexual flowers, with petals, tepals or floral receptacles forming a bowl shape, have been considered as distinctive features of the beetle pollination syndrome [22]. Within the Magnoliaceae, specific traits include the development of female floral structures before the male ones (protogyny) to ensure cross pollination, floral odors, floral movements and the production of heat by reproductive structures as a result of biochemical reactions (thermogenesis) [23].

Coleoptera (Scarabaeidae) have been observed feeding and mating in flowers of *Magnolia pugana* and *M. pacifica* (*M. talpana*), causing damage to the petals and sepals (**Figure 2**). Despite cantharophily is the most common pollination syndrome in basal angiosperms, hymenopterans have been observed visiting flowers of *M. pacifica* s. s. (Huajicori group) and *M. vallartensis*. In the latter species we can confirm that hymenopterans play a role in pollination given that we were able to photograph some bees with significant amounts of pollen attached to the specialized hairs in their hind legs (**Figure 2**). The known floral visitors to each of the studied *Magnolia* species are listed in **Table 1**. Insect pollination represents a fundamental ecosystem service in natural habitats, particularly for obligate insect-pollinated plant species [26, 27], however, there is still little knowledge on insect floral visitors and the floral biology of most of *Magnolia* species, therefore, further studies including more field observations are much needed.

2.2.3 Fruits and seeds

The fruits of *Magnolia,* often called polyfollicles, consist of spirally arranged carpels, initially free and in some species becoming concrescent. Dehiscence in all three species of the *M. pacifica* complex is dorsal and bivalve. *Magnolia granbarrancae* and the Huajicori group have ovoid and smaller fruits (**Figure 3**). Basal carpels are larger in *M. pacifica* than in any other species, while the carpel apex are distinctively more pronounced in *M. vallartensis*.

Magnolia seeds are attached to the fruit by thin funicular threads that keep them suspended after release for one or two days before falling to the ground. They have a three-layer coat, the outer one called sarcotesta is bright, red to orange colored (that is attractive for their dispersers, the birds), fleshy and with persistent odor oils

Figure 2.
Known floral visitors to the magnolias in the Magnolia pacifica *species complex. A–C,* Strigoderma sulcipennis *on* M. pugana*; damage to the petals and sepals by the beetle is visible. D–F,* Cyclocephala *sp. sheltering inside flowers of* M. talpana*. G–I, representatives of* Frieseomelitta nigra *visiting stamens of* M. vallartensis*; pollen can be seen attached to their hind legs. Photographs A–C by G. Hernández-Vera, D–I by M. Muñiz-Castro.*

	Taxonomic rank of floral visitors				
	Family	**Subfamily**	**Tribe**	**Genus**	**Species**
Magnolia pugana	Scarabaeidae	Rutelinae	Anomalini	*Strigoderma*	*Strigoderma sulcipennis* Burmeister [24]
Magnolia vallartensis	Apidae	Apinae	Meliponini	*Frieseomelitta*	*F. nigra* Cresson [25]
Magnolia talpana	Scarabaeidae Scarabaeidae	Rutelinae Dynastinae	Anomalini Cyclocephalini	*Strigoderma* *Cyclocephala*	x x
Magnolia pacifica (Huajicori group)	Apidae	Apinae	x	x	x

Table 1.
Known insect floral visitors to each of the studied Magnolia *species. The "x" indicates unidentified taxonomic rank.*

that make it impervious to water and gas exchange, containing inhibitors for germination and can protect it from some diseases and predators; the middle cover is called testa and provides mechanical protection as it is hard and bony, with a dark or light-colored depending on the species; the inner coat is a thin membrane that surrounds the endosperm, within which a rudimentary embryo is present [28, 29]. For the magnolias of the pacific complex, the seeds are prismatic to triangular with scarlet red sarcotesta and light-colored testa, presenting more similarity between *M. pugana* and *M. vallartensis* with seeds of the same size (9–12 × 7–8 mm) and shape (sub-cylindrical or round-compressed) and small variations with *M. pacifica* seeds, that are slightly bigger (8–14 × 7–11 mm) and sub-cylindrical to cylindrical or rounded to compressed [13].

2.2.4 Two new species of Magnolia from Jalisco

Magnolia granbarrancae A. Vázquez, Muñiz-Castro & A. Nuño sp. nov. (**Figure 3**, sixth column).

Type: MEXICO. Jalisco: Zapotlanejo municipality, on a slope on the southern margin of the Río Verde river, beside a rocky spring, 80 m NE from Las Cruces ravine, 3–5 km NNW from Matatlán, 20°44′30.4" N, 103°09′56.8" W, 1073 m, 23 Jan 2012 (fl), *M. Á. Muñiz-Castro, R. Murguía, J. Padilla-Lepe & M. Cházaro-Basañez* 918 (holotype IBUG!, isotype, MEXU).

Diagnosis. Magnolia granbarrancae is similar to *M. pugana* in having broadly obovate sepals, but it differs from the latter in having flowers smaller, 7.0–8.0 vs. 11–14 cm; a tight pollination chamber vs. a loose one; fruit smaller 3.9–5.3 vs. 5.0–7.5. cm, subglobose to broadly obovoid vs. oblongoid to ellipsoid; carpels more numerous 24–32 vs. 16–22, seeds orange vs. scarlet red.

Trees of (8) 10.0–15.0 m tall, 60–70 cm dbh, *leaves* 11.2–16.6 × (3.5) 4.6–6.3 cm, oblanceolate to elliptical, obtuse at the apex, acute at the base, glabrous, with 20 pairs of lateral veins; linear stipules 0.95–1 × 0.15–0.2 cm; petioles glabrous 1.7–2.0 × 0.1–0.15 cm, open *flowers* 10.0–11.0 cm in diameter, white; glabrous

Figure 3.
Morphological divergence in flowers and fruits of Magnolia *sect.* Magnolia *in Jalisco, Mexico.*

peduncles consisting of two internodes, 1.1–1.8 cm long, and 2.8–3.0 cm long; sepals 3, broadly obovate and strongly navicular, 5.8–6 × 3.1–3.5 cm, greenish beneath, glabrous on both sides; petals 6, at least, broadly obovate and strongly navicular, abruptly attenuated, the 3 outer ones 4.5–5 × 6–6.1 cm, white, glabrous in both faces, with a closed pollination chamber involving inner petals only 5.7–6.0 × 2.6–3.3 cm; stamens 40–46, *fruits* subglobose to broadly obovoid, glabrous, with 16–22 carpels, these, when mature and open, with almost round oval loci, 1.5–2.0 × 1.2–1.3 cm. *Seeds* broadly to narrowly ovoid or prismatic triangular or 0.9–1.1 × 0.6–0.9 cm, with orange sarcotesta.

Ethymology: the species was named after the great canyon in the vicinity of Guadalajara, as Henri Galeotii once name it: "la grande Barranca de Guadalaxara" (McVaugh 1952, Asa Gray Bulletin, Ann Arbor Michigan).

Distribution and habitat: Endemic to the Río Verde canyon, in the municipalities of Zapotlanejo and Ixtlahuacán del Río, Jalisco. Inhabits in steep slopes between 1073 and 1215 m asl, on rocky springs surrounded by tropical dry forest and some mesophytic plants as *Oreopanax peltatus, Trema micrantha, Ardisia sp., Aphananthe monoica, Piper sp.*

Additional specimens examined. JALISCO. Zapotlanejo municipality: Las Cruces ravine, 30 m down the stream pond and the water pump, 20°44′22.21" N, 103° 9′48.01" W, 1215 m, 23 Jan 2012 (fl), *Muñiz-Castro, et al.* 919 (IBUG!). Ixtlahuacán del Río Municipality: North slope of the Barranca del Río Verde, next to a spring, in a mango orchard, 80 m above the river, 1.2 km ENE upriver of the Atengo canyon, 20° 44′0.33" N, 103°11′23" W, 1204 m, 23 Feb 2013 (fr), *Muñiz-Castro, et al.* 1161 (IBUG!,)

Magnolia talpana A. Vázquez, Muñiz-Castro & S. Ortega sp. nov. (**Figure 3**, third column).

Type: MEXICO. Jalisco: Vicinity of Parque Estatal Bosque de Arce. Municipio Talpa de Allende, 2.1 km SW from Los Sauces 20°14′42"N, 104°47′41"W, 1340 m, 12 Abr 2012 (fr), riparian cloud forest, besides a small tributary stream of the Talpa river. *M. Á. Muñiz-Castro & R. Murguía* 970 (holotype: IBUG, isotype: IPN!).

Diagnosis. Magnolia talpana is similar to *M. pacifica* in sharing ellipsoid fruits with narrow locules during dehiscence, but it differs from the latter in having flowers smaller, 10.0–11.6 vs. 12–14 cm; a tight pollination chamber involving the outer and inner whorl of petals vs. a loose pollination chamber involving the inner whorl of petals only; open sepals narrowly oblongoid, reflexed and opening up to 170 degrees vs. oblongoid, not reflexed and opening less than 90 degrees; inner whorl of petals varying in size vs. subequal; peduncles pubescent vs. glabrous.

Trees of 15.0–25.0 m tall, *leaves* 8.1–14.9 × 5–7.5 cm, acute at the base, acute to obtuse at the apex, glabrous, with 14–15 pairs of lateral veins; stipules linear or conic when young, and abaxially golden pubescent, yellowish green; petioles glabrous, open *flowers* 10–11.6 cm in diameter, white; a tight pollination chamber involving the outer and inner of petals, 4.8 × 2.1–2.7 cm; peduncles of five internodes, the longest is the most distal one, half the length of sepals (ca. 2.5 cm long); sepals 3, narrowly oblongoid, navicular reflexed and opening up to 170 degrees; 4.3–5.5 × 1.4–2 cm, white, glabrous on both sides; petals 6, obovate, navicular, abruptly attenuated, the 3 outer ones 5.2–5.9 × 2.9–3.6 cm, white, glabrous in both faces, the 3 inner whorl of petals subequal, 5.3–6.1 × 2.5–2.8 cm; stamens 74–76; gynoecium 2.1 × 1.0 cm, *fruits* ellipsoid with narrow locules during dehiscence, immature fruit 2.0 × 1.1 cm. *Seeds* with red sarcotesta.

Distribution and habitat: Endemic to Talpa de Allende municipality, Jalisco, México, in the east and west branches of the high watershed of the Talpa river and in the Camacho and Desmoronado tributaries of the Tomatlán river, north and above Presa Cajón de Peñas. Inhabiting in cloud forest, riparian forests and ecotones with oak-pine forest.

Additional specimens examined: MEXICO. Jalisco: Municipio Talpa de Allende, 120 m SSW from Peña del Cuervo, 20°13′4.1”N, 104°44′11”W, 2077 m, 16 Jun 2012 (fl), *Muñiz-Castro and Murguía-Araiza 1094* (IBUG!); Cañada Ojo de Agua del Cuervo (Maple Forest), 18.2 km SSE from Talpa de Allende town, 1751 m, 01 Aug 2012 (fl) *Muñiz-Castro et al. 1106* (IBUG!); 17–19 km S from Talpa de Allende, margin W of Río de Talpa, 18–19 Oct 1960 (fr), *McVaugh 20375* (MICH!); 20–22 km S from Talpa de Allende, margin W of Río de Talpa, 28–30 Mar 1965 (fr), *McVaugh 23313* (MICH!); Cañada Ojo de Agua del Cuervo (Maple Forest), 200 m from the road, 1754 m, 24 Mar 2012 (fr), 20°12′46.5”N, 104°45′25.4”W, *Muñiz-Castro et al.* 941 (IBUG!); path to Coamil de Méndez, tributary of Río Talpa, 2.5 km SW from Los Sauces, 1378 m, 31 Jul 2012 (fr), 20°14′19.1”N, 104°47′48.1”O, *Muñiz-Castro*, *et al*. 1103 (IBUG!).

2.2.5 Key to the species of Magnolia in Jalisco.

1. Stipules adnate to petiole; fruits with connate carpels, circumcissil dehiscence and detachable mainly singly or in small irregular groups (sect. *Talauma*) ... 2.

    - Free petiole stipules, free carpels, fruit with dorsal dehiscence (sect. *Magnolia*)

2. Leaf blades 24.0–25.0 × 11.9–12.6 cm; flowers 22.0 cm diameter, fruits 7.0–10.0 × 5.0–7.0 cm; carpels 47–58; basal carpels 4.3–4.5 × 1.2–1.4 cm and their decurrence from 0.4–0.6 cm long (S of Jalisco and Colima) ... *M. jaliscana*

    - Leaf blades 35.0–45.0 × 23.0–29.0 cm; flowers 16.0 cm in diameter; fruits 14.5 × 8.5–9.0 cm; carpels 37–44; basal carpels 5–5.7 × 1.5–2 cm and their decurrence from 0.8–2.0 cm long (O of Jalisco) ... *M. ofeliae*

3. Bract, peduncular internodes and petioles densely pubescent ... *M. iltisiana*

    - Spataceae bract, peduncular internodes and petioles essentially glabrous or with pubescence limited to the nodes ... 4.

4. Widely obovate sepals (N of Jalisco and S of Zacatecas) ... 5.

    - Narrowly oblong sepals ... 6.

5. Flowers 7.0–8.0 cm in diameter, with a tight pollination chamber; fruit 3.9–5.3 cm, subglobose to broadly obovoid; carpels 24–32, seeds orange ... *M. granbarrancae*

    - Flowers 11–14 cm in diameter, with a loose pollination chamber, fruit 5.0–7.5, oblongoid to ellipsoid, carpels 16–22, with seeds scarlet red ... *M. pugana*

6. Pollination chamber subglobose to globose, leaves 13.5–27.8 × 6–14.8 cm, broadly elliptic to elliptic, often obtuse to rounded apex, petals 6–8, carpels 10–19 ... *M. vallartensis*

    - Pollination chamber incipient or narrowly oblongoid, leaves 8.0–17.0 (18.0) × 3.0–6.0 (8.0) cm, elliptical-lanceolate, frequently acute apex, petals 6–7, carpels 17–25 ... 7.

7. Sepals oblongoid, not reflexed, opening less than 90 degrees, pollination chamber loose and incipient, inner whorl of petals of greatly varying in size ... *M. pacifica*

   - Sepals narrowly oblongoid, reflexed, opening up to 170 degrees, pollination chamber tight and narrowly oblongoid, inner whorl of petals subequal ... *M. talpana*.

2.3 Genetic structure, diversity, and differentiation

In this section, we review what is known so far of genetic structure and diversity in three western Mexican *Magnolia* species: *Magnolia vallartensis, M. pacifica* s.l. (including the new species *M. talpana*), and *M. pugana* s.l. (including the new species *M. granbarrancae*), all belonging to the so-called *Magnolia pacifica* complex [11] (**Figure 4**). As they are distributed along a continentality and moisture gradient, the influences of their different distances to the Pacific Ocean in their genetic structure and diversity have been tested [18]. In this study, leaf tissues of 278 individuals from 10 localities representative of the *M. pacifica* complex were sampled. Three localities were sampled for *M. vallartensis*: Arroyo Palo María (APM), Las Lajitas (LL), and Provincia (PV); three for *M. pacifica* s.l.: Cerro San Juan (CSJ), San Sebastián del Oeste (SS), and Bosque de Arce (BA, *M. talpana*); and four for *M. pugana* s.l. Arroyo La Virgen (ALV), Arroyo San Lorenzo (ASL), Arroyo Palo

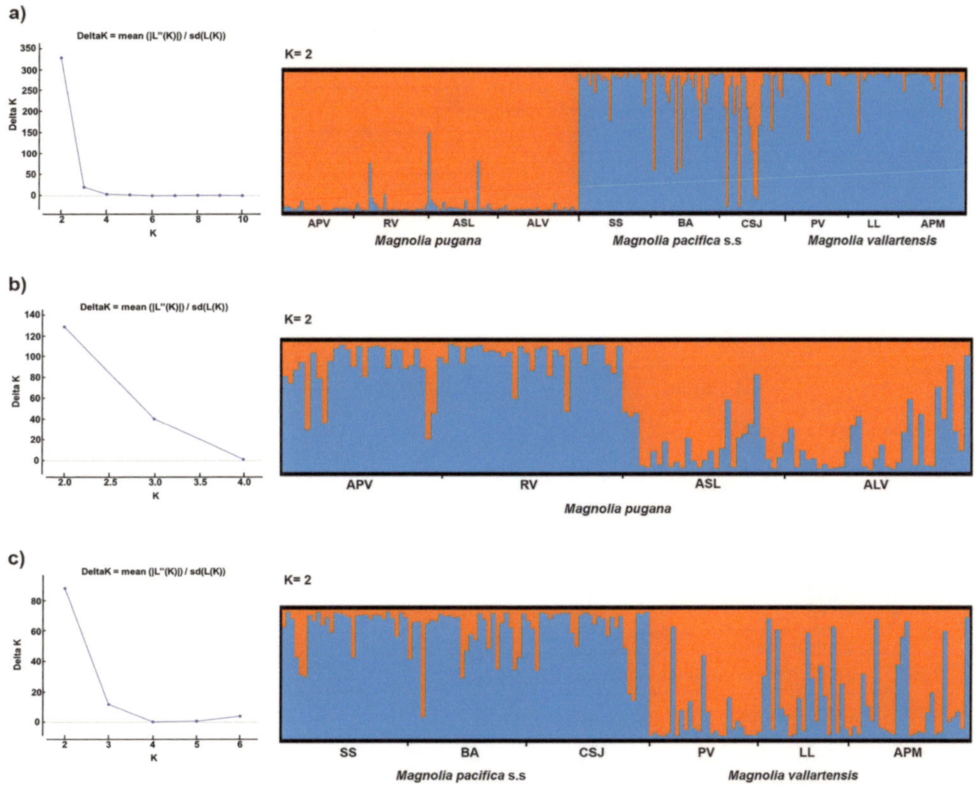

Figure 4.
Results of Bayesian clustering based on STRUCTURE; analysis performed on a) the entire data set (278 individuals, 76 loci ISSR); b) the M. pugana *s.l. group (120 individuals); c) the* M. pacifica *s.l.-*M. vallartensis *group (158 individuals). Relationships between K and Delta K values and structure bar plots are shown. Vertical bars represent each individual analyzed and bars are divided into distinct colors when there is evidence of admixture. From [18].*

Verde (APV), and Río Verde (RV, *M. granbarrancae*) (**Figure 4**). A total of 76 reproducible DNA fragments of six Inter-Simple Sequence Repeats (ISSR) (primers UBC 810, 814, 834, 836, 855, and 857) were amplified and analyzed using Bayesian analysis with the program STRUCTURE, UPGMA clustering, Monmonier's algorithm with Barrier 2.2, AMOVA, and genetic diversity parameters. Also, a Mantel test was done to evaluate the relationship between genetic and geographic distances.

2.3.1 Genetic structure

Results of Bayesian analysis with STRUCTURE (**Figure 4**), UPGMA clustering (**Figure 5**), and the Exact Test for differentiation (**Table 2**) are in accordance that there are two main genetic clusters for the whole of three species, being *M. pugana* s.l. the eastern main group, and the cluster of *M. pacifica* s.l.-*M. vallartensis* the

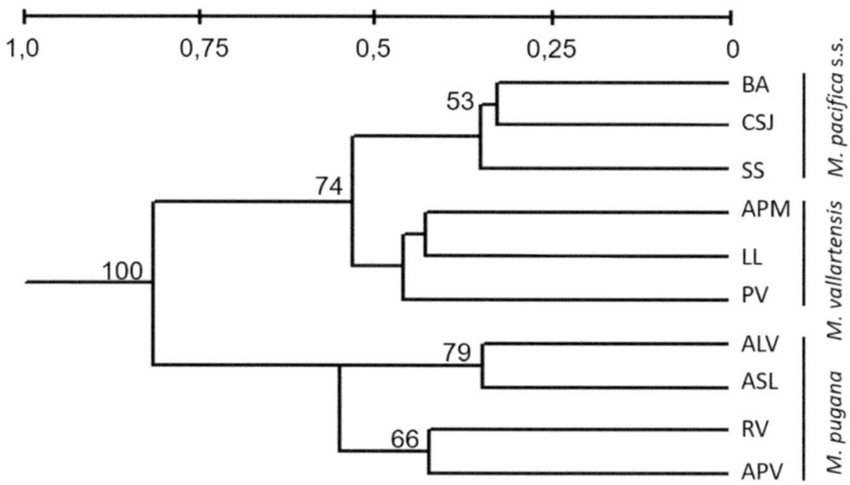

Figure 5.
UPGMA dendrogram showing the genetic relationships of the M. pacifica *species complex, the number at each node indicates the bootstrap percent having above 50% values. From [18].*

	M. pugana				*M. pacifica* s.s.			*M. vallartensis*		
	APV	**RV**	**ASL**	**ALV**	**SS**	**BA**	**CSJ**	**PV**	**LL**	**APM**
APV	—	0.039	0.051	0.050	0.069	0.057	0.060	0.098	0.097	0.088
RV	0.079	—	0.054	0.052	0.074	0.070	0.073	0.110	0.112	0.088
ASL	0.006	0.001	—	0.032	0.060	0.065	0.050	0.077	0.079	0.072
ALV	0.013	0	0.144	—	0.070	0.069	0.066	0.099	0.097	0.080
SS	0	0	0.002	0	—	0.031	0.030	0.058	0.042	0.049
BA	0.002	0	0	0	0.520	—	0.032	0.065	0.042	0.045
CSJ	0	0	0	0	0.312	0.703	—	0.056	0.042	0.040
PV	0	0	0	0	0.060	0.052	0.050	—	0.039	0.043
LL	0	0	0	0	0.060	0.475	0.119	0.976	—	0.038
APM	0	0	0	0	0.360	0.210	0.162	0.550	0.360	—

Table 2.
Nei's unbiased genetic distance (above diagonal) and exact test differentiation probability values (below diagonal) among sampled localities of the M. pacifica *species complex. From [18].*

western grouping. But each one of these two main groups has its own structure, having two subgroups each (**Figures 4** and **5**). *M. pugana* s.l. group is segregated in the *M. pugana* s.s. subgroup (ALV-ASL localities), located SW of the Santiago river canyon, and the APV-*M. granbarrancae* (RV) group, located NE of that canyon. In the other main group, *M. pacifica* s.l. subgroup (BA-SS-CSJ) was separated from the *M. vallartensis* one (PV-LL-APM). Additionally, the 900 bp (primer 834) and 850 bp (primer 855) ISSR loci only amplified and were exclusive for *M. pacifica* s.l.-*M. vallartensis* group, these loci did not amplify *M. pugana* s.l. samples. The AMOVA based on the three species and the two Bayesian groups showed that only 8% of the genetic variation was explained by differences among species as well as among the Bayesian groups (**Table 3**), so AMOVA was less informative than the Bayesian analysis. Most of the variation ($\geq$ 82%) was explained by differences within localities, suggesting high levels of cross-pollination [30]. This predominance of outcrossing reproductive mode was also found by [31] in some Neotropical magnolias.

Other evidence for the structure represented by four groups was the results of the test with Monmonier's algorithm (Barrier 2.2), which detected three significant geographical barriers to gene flow, segregating the four groups (**Figure 4**). All geographical boundaries had 100% bootstrap support. One of these is the Trans-Mexican Volcanic Belt (TMVB), which is the main physiographic barrier between *M. pugana* s.l. and the group of *M. pacifica* s.l.-*M. vallartensis*. This mountain range has been identified as an important gene flow barrier for both plants [32–35] and animals [36, 37], even its complex orogenic processes have led to population isolation, speciation, and diversification [38]. The other two significant barriers detected correspond to the Santiago river canyon on the one hand (between the *M. pugana* s. s. and the APV-*M. granbarrancae* group), and the basins of the rivers Ameca, Mascota, Pitillal, and Cuale on the other hand (between *M. pacifica* s.l. and *M. vallartensis*). The effect of the deep Santiago river canyon (3–15 km wide and 500–700 m deep) and other river basins as significant gene flow barriers also have been documented for birds, reptiles, insects and other plant species [34, 39, 40].

The outcomes of this population genetics study reveal that *M. pugana* s.l., the easternmost and the most continental group, is the most geographically and genetically distinct group of all *M. pacifica* species complex, and clearly support its recognition at the species level [16]. The most likely process of speciation for *M. pugana* s.s. and *M. granbarrancae* is allopatric isolation, being the TMVB the main

	Percentage of variation				
	Groups		Taxa		
Variation source	**Bayesian analysis (2)**	**Taxonomic (3)**	***M. pugana* s.l.**	***M. pacifica* s.l**	***M. vallartensis***
Among					
Taxa / Genetic groups	9 (0.001)	8 (0.001)			
Localities	9 (0.001)	8 (0.001)	12 (0.001)	7 (0.001)	7 (0.001)
Within					
Localities	82 (0.001)	84 (0.001)	88 (0.001)	93 (0.001)	93 (0.001)

(2) Two Bayesian groups, (3) three taxa, p value is given in parentheses.

Table 3.
Molecular variance analysis (AMOVA) for Bayesian analysis and taxonomic groups of the Magnolia pacifica *species complex. From [18].*

gene flow barrier between them and their western close relatives. This is in congruence with Vázquez-García *et al.* [1], who suggest that allopatric speciation is a major driver of *Magnolia* diversification in Neotropical Magnoliaceae. The genetic divergence between *M. pacifica* s.l. and *M. vallartensis* is lesser than between this main group and *M. pugana* s.l., indicating a more recent process of segregation or current gene flow. There is no clear geomorphological barrier between some localities of *M. talpana* and *M. vallartensis* in their southern ranges (between BA and all the *M. vallartensis* localities). Therefore, genetic differences between these localities might be rather explained by a process of parapatric ecological differentiation [41, 42], as it is common that gene exchange among closely related taxa happens in at least 25% of plant species [43]. Nevertheless, the four identified genetic groups, independently of partially sharing genes, are distinctive evolutive entities and should be considered as separated conservation units.

2.3.2 Genetic differentiation

Differentiation indices were moderate in general, but higher in *M. pugana* s.l. (GST = 0.120 ± 0.021, D = 0.028 ± 0.007) than in *M. pacifica* s.l.–*M. vallartensis* (GST = 0.106 ± 0.016, D = 0.026 ± 0.006). The Exact Test for population differentiation showed significant genetic differences between localities of *M. pugana* s.s. and the *M. granbarrancae*, and between localities of *M. pugana* s.l. and *M. pacifica* s. l.–*M. vallartensis*, but not within each *M. pugana* subgroup and within the *M. pacifica* s.l..–*M. vallartensis* group (**Table 2**). The higher genetic differentiation of *M. pugana* s.l. is in accordance with its smaller, more fragmented, and more isolated populations. The highest distance to the Pacific Ocean of *M. pugana* s.l. causes a lower environmental humidity and more extreme cyclical temperature changes, that is, a greater continentality [44]. These drier and more extreme conditions result in declines, fragmentation, and isolation of *M. pugana* s.l. populations, which need a constant input of water to survive. In contrast, the more humid maritime environments of *M. pacifica* s.l.–*M. vallartensis* are more favorable to maintain populations of these cloud forest mesophytic species, which is reflected in less isolation and differentiation. Similarly, genetic differentiation has ranged from moderate to high in most Neotropical Magnolia species [31].

2.3.3 Genetic diversity:

M. pugana s.l. had lower genetic diversity than the *M. pacifica* s.l.-*M. vallartensis* group. *M. pugana* s.l. exhibited a Shannon Index (I) = 0.268, total heterozygosity (H_T) = 0.158 (0.023 SD), and intrapopulation heterozygosity (H_S) = 0.134 (0.020 SD). In contrast, *M. pacifica* had an I = 0.272, H_T = 0.175 (0.025), and H_S = 0.159 (0.023), and *M. vallartensis* an I = 0.275, H_T = 0.171 (0.024) and H_S = 0.153 (0.022). Genetic diversity also varied among localities; *M. pugana* s.l. localities had the lowest genetic diversity (H_E = 0.121–0.140, *M. granbarrancae* having the lowest I = 0.218 and H = 0.121), whereas the locality Bosque de Arce (BA) of *M. talpana* showed the highest genetic diversity (I = 0.280, H_E = 0.17) among the western *Magnolia* localities studied. The genetic diversity estimates for all the studied taxa here were lower than the average values reported for plant genetic diversity based on ISSR (H = 0.22) [45], and much lower than those reported for two threatened eastern Mexican *Magnolia* species: *M. sharpii* Miranda [46] (I = 0.56) and *M. schiedeana* Schlecht. [47] (I = 0.50) [48]. As with the higher genetic differentiation, the lowest genetic diversity of *M. pugana* s.s. and *M. granbarrancae* is consistent with their smaller populations, higher isolation, and fragmentation, all influenced by a drier and more extreme climate.

2.3.4 Isolation by distance pattern

The correlation between geographical and genetic distances among all localities of the *M. pacifica* species complex, revealed by Mantel tests, was high and significant ($r = 0.80$, $p < 0.001$), however, when the tests were applied to each of the two main genetic groups separately, no significant correlation was detected. Isolation by distance might explain the genetic structure and differentiation pattern within the *M. pacifica* species complex. Isolation by distance has been a strong pattern observed among plant studies, under this and in rapid environmental change, adaptive responses to environmental stress will be constrained by the natural dispersal mechanisms [49]. Pollination by beetles in small and isolated populations of *Magnolia* is not very efficient [50], and habitat fragmentation and other anthropogenic factors may also be troublesome for *Magnolia* seed dispersal by birds [50, 51]. Even if dispersal occurs, seeds may not germinate or seedlings may not survive in places without enough humidity [52].

In summary, based on ISSR genetic variation, the *M. pacifica* species complex exhibits a population genetic structure composed of two main groups, the eastern *M. pugana* s.l more continental group, and the western *M. pacifica* s.l.-*M. vallartensis* group, with the more maritime climate. Both main groups are segregated by the physiographic barrier of the TMVB and isolation by distance, and are at the extremes of a maritime-continental climatic gradient. *M. pugana* s.s. and *M. granbarrancae* are subject to a drier and more extreme climate, therefore having more deforested, fragmented, and isolated habitats, which leads to lower genetic diversity and a higher genetic differentiation. This differentiation within *M. pugana* s.l. and the physiographic barrier of the Santiago river canyon have structured this taxa in two genetic subgroups, *M. pugana* s.s. and the *M. granbarrancae*. The *M. pacifica* s.l-*M. vallartensis* group exhibited genetic segregation in two subgroups, having several canyon river barriers between both taxa, but maintaining a partial gene flow at their southern ranges. The three species of the *M. pacifica* complex have lower genetic diversity than eastern Mexican *Magnolia* species which are considered as endangered. Even more, the eastern and more continental *M. pugana* s.s. and *M. granbarrancae* undergo the lowest genetic diversity, which, together with their smaller and more isolated populations, makes these populations more vulnerable to gene drift and bottlenecks, therefore greater risk of extinction. All main genetic groups and subgroups defined in this study should be considered as separate conservation units, and concerted efforts are needed to protect them.

2.4 Phenological divergence

Phenology aims to characterize the behavior and the biological adaptation of the species in its natural habitats, which may help guide important management conservation strategies. We studied three different species of *Magnolia* (*M. vallartensis*, *M. pacifica* s.l., *M. pugana* s.l.) in western Mexico (**Figure 1**). Phenological observations were made at monthly intervals for one year and four months. The phenophases (flowering and fruiting) in an annual cycle and the intensity of each phenological event were estimated using the Fournier intensity index. The sum of all the intensity categories assigned to each individual was divided by the maximum amount that could be attributed to the population (relativizing the total number of selected individuals with the total sum sampled per month). Circular statistics were used to analyze and interpret phenological patterns. Using the statistical package Oriana version 4.0, the months were converted into angles, with intervals of 0.999° to 30° wide and it was calculated: 1) average angle (μ) or average date that refers to the time of the year around which the phenological activity of individuals is more concentrated; 2) circular standard deviation; and 3) the vector r, which indicates the intensity of the concentration (0 to 1) around the average angle. The length of the vector r can be considered a measure of the degree of seasonality. In addition, the Rayleigh test was applied to indicate the significance ($P < 0.05$) of the mean angle.

The synchrony of the flowering phenophase between populations or species is essential to detect possibilities of gene flow between species and existence of temporary barriers in reproductive phenophases. Since the incipient asynchrony observed in the flowering of the three species was linked to their amplitude of flowering (which lasts up to six weeks with significant intersection) it was considered insufficient to prevent gene flow. The hypothesis of non-seasonality in the phenophases of the three species was rejected, which was high in *M. pugana*, intermediate in *M. pacifica* and low in *M. vallartensis*.

The flowering (female-flower) phenophase showed an incipient asynchrony among species, being earlier in *Magnolia vallartensis* (end of May), intermediate in *M. pacifica* (second week of June), and late in *M. pugana* (end of June). Seasonality was high in *M. pacifica* and *M. pugana* [20] (**Figure 6**). This incipient asynchronous female-flower phase differs from some Colombian *Magnolia* species, in which flowering is commonly markedly asynchronous among species and populations of each species [53].

The fruiting phase was asynchronous among species, and showed high seasonality only for *Magnolia pugana* (February) while in *M. vallartensis* the maximum fruiting was in August and for *M. pacifica* in December (**Figure 6**). This pattern is similar to that reported for *Leucaena* sp. and *Guazuma ulmifolia* Lam. [54, 55], a possible strategy to reduce competition among potential seed dispersers of *Magnolia* including birds, rodents and lizards. The intensity of fruiting is concentrated in rainy seasons [53]. Fruit dehiscence of *Magnolia vallartensis*, *M. pacifica* and *M. pugana* showed an asynchronous pattern between species. Similarly, eight

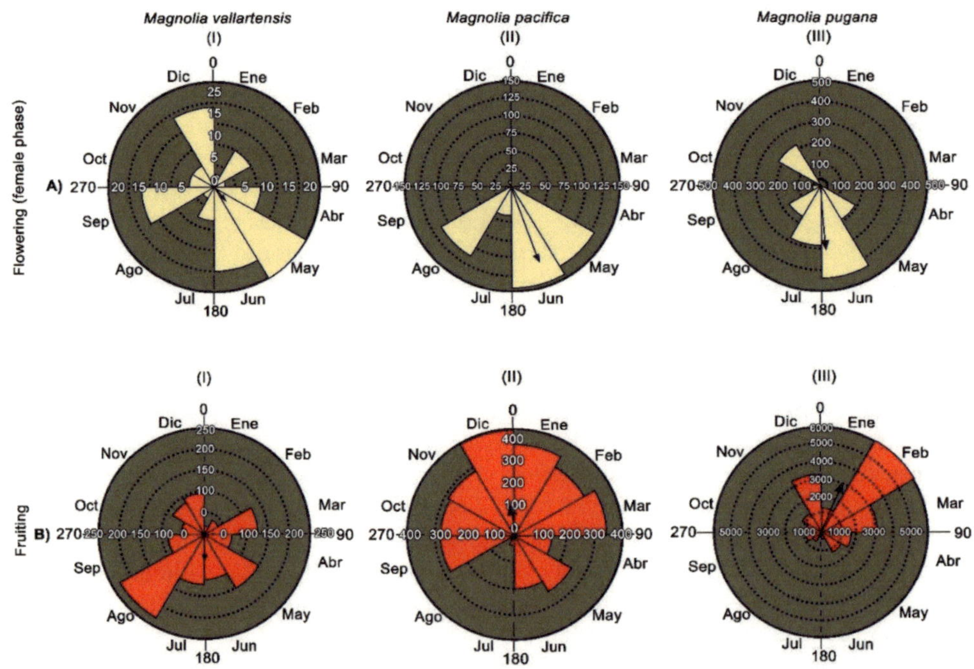

Figure 6.
Circular histograms of two phenophases for the three Magnolia *species. A. Flowering (female-flower) phenophase. B. Fruiting phenophase. Based on monthly averages of frequencies from January to December 2017.* Magnolia vallartensis *(I), left column,* M. pacifica *(II), central column and* M. pugana *(III), right column. The arrows represent significant average angles (Rayleigh test $p < 0.001$), while the length of the vector r (0 to 1) indicates the concentration around the average angle [20].*

species with an asynchronous pattern are reported for pioneer species in the central Amazon region of Brazil [56].

Differences in productivity were also evident, the higher productivity of reproductive phenophases (flowering and fruiting) of *Magnolia pugana* compared to that of *M. pacifica* and *M. vallartensis* makes it possible to reject the hypothesis of no significant differences. These differences may be due to their evident genetic differences [18], but also to their environmental differences; for instance, *M. pugana* populations thrive in permanent streams, surrounded by tropical deciduous forest and open oak-pine woodlands, a condition that provides more sunlight that can help to keep their reproductive phenophases more productive (flowers and fruits), whereas in the case of *M. vallartensis* and *M. pacifica*, their populations are usually surrounded by high and dense canopy cloud forest and tropical semi-evergreen forest trees that overshadow many *Magnolia* trees, which limits their productivity due to the scarcity of sunlight along the year.

2.5 Floral scents

More recently, several ecological studies have examined the roles of floral scent in the biology of the plant [57]. Scents, essential oils, or volatiles are secondary metabolites produced by plants to fulfill protection functions against herbivores, phytopathogens, and even other plants [58, 59]. These form a complex matrix with a lipophilic-volatile nature [60], from a chemical point of view, these volatile compounds belong to chemical classes such as terpene derivatives (oxygenated or hydrocarbons), phenylpropanoids, benzenoids, and nitrogen-containing compounds [61, 62]. The combinations of the constituents of this scent mixture give each flowering plant species a unique fragrance [57]. Through a compilation of data, the genus *Magnolia*, has been reported to be an interesting source of secondary metabolites found both in plant extracts and in essential oils, as shown by studies carried out mainly in species distributed in Asia and in which applications in traditional medicine are included [63–65]. In this sense, in many of the endemic species, studies related to describing the metabolic composition as a complement to determine potential uses, contribute to knowledge, and improve the conservation of their habitats. As an example, we extracted and characterized for the first time the essential oils present in flowers of *M. vallartensis*, *M. pacifica* and *M. pugana*, three endemic species of the genus *Magnolia* along a continenrtality and moisture gradient in western Mexico [19].

2.5.1 Methods

For details of flower collection techniques, essential oils extraction and determination of floran scents chemical composition see Mendeley Data repository [66].

2.5.2 Comparisons

The yield of essential oils from flowers of these three *Magnolia spp.* (**Table 4**) showed that in general these vegetal parts present a similar low yield between 0.2 to

Specie	Flower mass (g)	Essential oil (mL)	Yield	Odor
M. pacifica	100	0.22	0.21%	Sweet, citric
M. vallartensis	100	0.24	0.25%	Sweet, citric
M. pugana	100	0.30	0.30%	Sweet, woody

Table 4.
Essential oils of Magnolia *flowers: Yield and organoleptic characteristics.*

0.3%. On the other hand, the chemical composition of oils showed qualitative differences among individual components. The chemical profiles of these scents analyzed by GC/MS, observed in **Figure 7**, identified 97 compounds in total between species; 63 in *M. vallartensis*, 53 in *M. pacifica* and 39 in *M. pugana*.

In the chromatograms differences evidenced, in a specific manner, on the major components. Within the chemical composition of floral essential oils the major compounds greater than 3% of the total components obtained for each species from the most abundant were as follows. In the case of *M. pacifica*, (caryophyllene, bicyclo-dec-1-ene, 2-isopropyl-5-methyl-9-methylene, bicyclogermacrene, β-elemene and epi-cyclocolorenone); for *M. vallartensis* (caryophyllene, geranyl methyl ether, β-elemene and caryophyllene oxide); and in *M. pugana* (cyclocolorenone; 2Z,6E-farnesol; benzoic acid (5,5-dimethyl-4-oxo-2-cyclohexenyl ester, β-elemene, and caryophyllene oxide). Chemical composition data has an interesting application when essential oils occurrence is studied in the field of taxonomy. This assesses the possible chemotaxonomic relationship between chemical compounds and species, identifying compounds that could act as indicators. Based on this idea, the results obtained by a point-to-point analysis throughout the 90-minutes chromatograms and comparing the mass spectra including trace components, a matrix was constructed (**Figure 8**).

The presence of compounds could be associated to the closely related taxonomic affinity confirmed trough molecular phylogenetic analysis; for example analyzing other three species closely related: *M. schiedeana*, *M. grandiflora* L. [67] and *M. tamaulipana* [11]; the pair *M. schiedeana-M. tamaulipana* (both endemic from Mexico) share the presence of geranyl methyl ether as the major compound up to 87%, while in *M. grandiflora* (widely distributed in the southern North America) shows the presence of its precursor: geraniol [68]. In this way, it was determined that the floral scents of the endemic species *M. pacifica, M. vallartensis* and *M. pugana* share qualitatively 14 components in their chemical profile, which is equivalent to 14.46% of total; these components suggested that could act as chemical markers for a determination at the gender level. Otherwise, floral oil compounds shared by pairs of species were also founded, the pair *M. pacifica-M. vallartensis* shared 27 more compounds of the total (42.47%); *M. vallartensis-M. pugana* 2 more compounds of the total (16.49%); and *M. pacifica-M. pugana* just one more additional compound to the total (15.46%). In the case of components particularly present in each species, *M. pugana* had 22 compounds, *M. vallartensis* 20 and

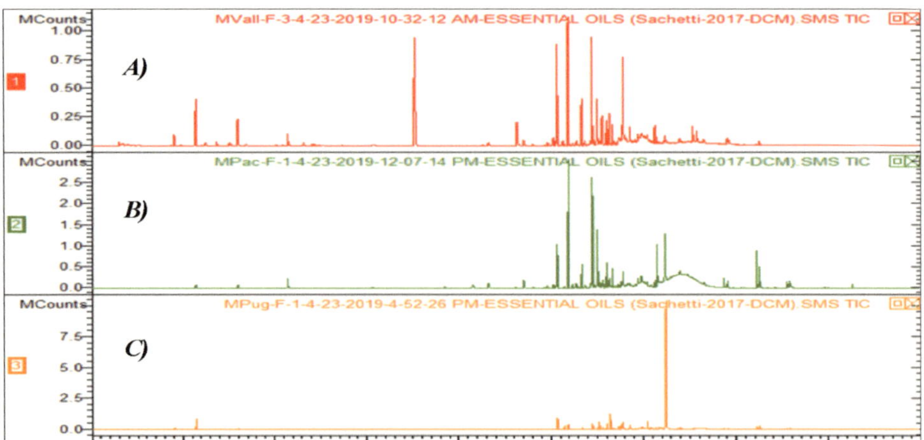

Figure 7.
Floral scents chemical profiles by gas chromatography: A) M. vallartensis, B) M. pacifica, *and* C) M. pugana.

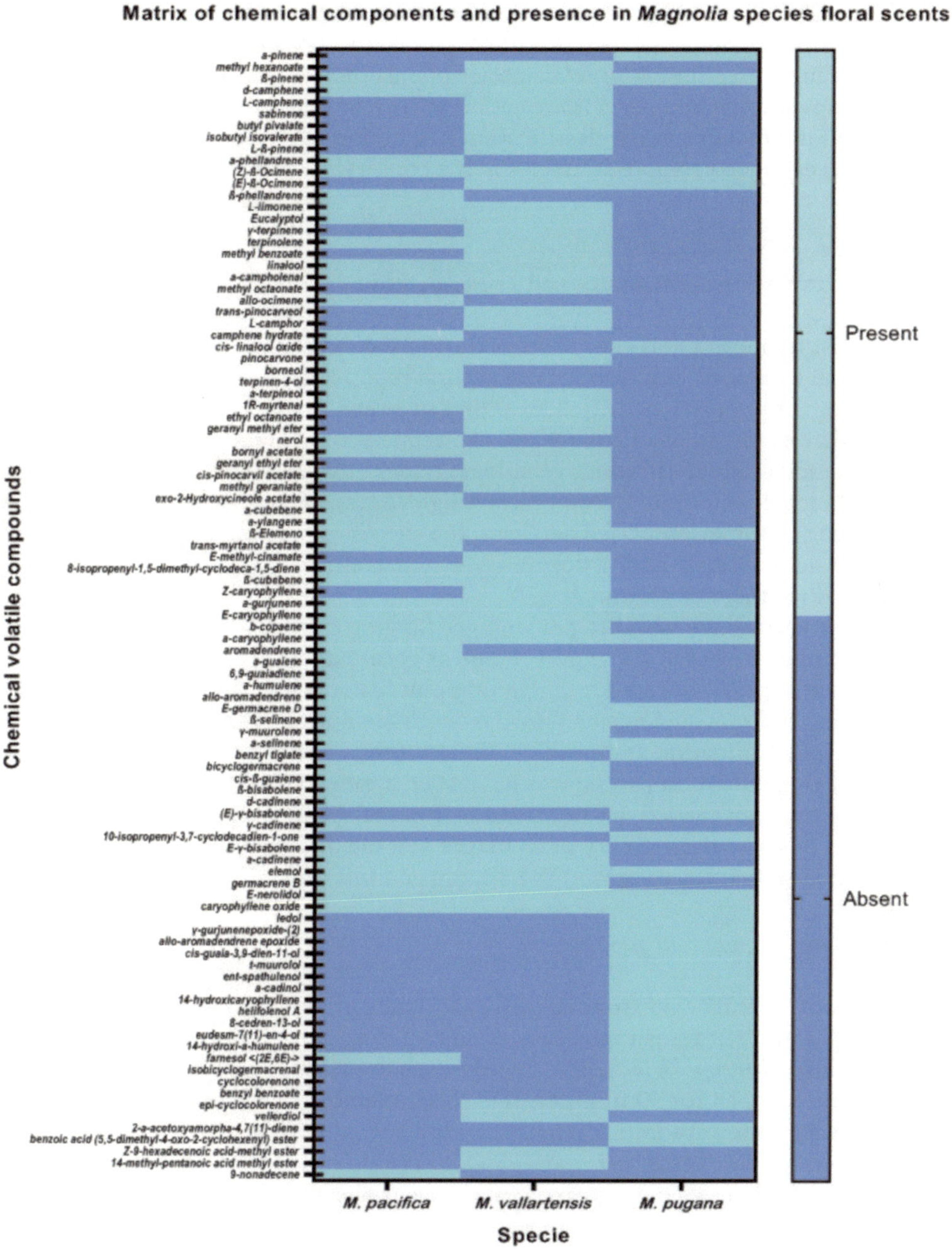

Figure 8.
Matrix of chemical compounds present in Magnolia *floral scents.*

M. pacifica 11 components. The Jaccard similarity index allowed inferring the qualitative composition, the pair *M. pacifica - M. vallartensis* show the strongest similarity about 54.6%, while with *M. pugana* these species showed 19.5% and 18.6% of similarity respectively. The essential oils of flowers of *M. pacifica*, *M. vallartensis* and *M. pugana* showed a composition rich in sesquiterpenes with significant differences in their composition. Most species of *Magnolia* possess distinctive floral scent profiles, even though they may be taxonomically closely related, the chemical differences among these taxa may have arisen from interaction with pollinators or the environment [69]. There are factors that can affect the chemical composition and quality of an essential oil such as the age of the plant, altitude, climate, genetics, geography, type of soil [70]. For this reason, it is necessary to continue with studies focused on establishing the relationships between the components and relating the chemotaxonomic field including more *Magnolia* species populations to conform a robust database that provides the necessary information to correlate chemical markers and to become a useful tool in species classification in addition to genetic analysis. These applications can help promote strategies in the conservation of this genus and its habitat.

2.6 Pre-germination treatments and seed dormancy

The most important and critical stage during the development of the plants is germination; therefore, it is also for the natural repopulation of the species and the dynamics of the populations [71]. The ability of a seed to germinate is known as viability, that is, the embryo is alive and can remain so for some time. A viable seed with dormancy is one that cannot germinate under any condition of physical environmental factors (humidity, temperature, light/dark, etc.), which are otherwise favorable for germination [72]. Therefore, the success of germination depends on whether or not the seeds have a dormancy mechanism, on their viability and on the time needed for them to germinate [73].

Magnolia seeds have viability percentages ranging from 50–100% [74, 75]. Different studies have shown a low percentage of germination because they can present different types of dormancy (including physical, chemical, mechanical, physiological and morphological) [76]. To eliminate dormancy and increase the percentage of germination and produce the greatest number of *Magnolia* seedlings, different pre-germination procedures have been applied [77–80]. This section will present the research carried out in *Magnolia iltisiana*, *M. pacifica*, *M. pugana* and *M. vallartensis*, to know the viability percentages and the different most successful pre-germination treatments, to break the lethargy and increase germination.

2.6.1 Seed management and germination

Seeds must be collected from 10 different plants of ripe fruit and extracted manually. To prevent fungal infection, a contact fungicide (Captan) was used for the seeds of *M. iltisiana* [81], while the other species were rinsed in a 3% sodium hypochlorite solution for 30 minutes. It is recommended that after being collected and disinfected, they are immediately stored in a well-ventilated and humid environment inside a refrigerator at 4–5°C, to avoid dehydration [76].

Were used 100 seeds per treatment with five replicates of 20 seeds per container for the four species. Once the treatments were concluded seeds of *M. pacifica*, *M. pugana* and *M. vallartensis* were sown inside a greenhouse at the Centro Universitario de Ciencias Biológicas y Agropecuarias (CUCBA) and were buried at 1.5 cm depth in 25 ml plastic containers. The substrate used for planting was "peat-moss" [29]. The study of *M. iltisiana* was carried out in the experimental greenhouse of Las Joyas Scientific Station (ECLJ), and they were planted in a mixture of sand and "germinaza" (1:1) as substrate. The number of germinated seeds was recorded daily for 60 days [75]. Seeds were considered to have germinated when the radicle emerged or when the hypocotyl was observed [82, 83].

2.6.2 Viability tests

The percentage of viability of all species was determined through tests in a 1% tetrazolium solution; two replicates of 50 seeds were used for *M. iltisiana*, while for the other species it was in 30 seeds. The seeds were immersed without aril in this solution for 24 hours at 30°C, in darkness [72]. After this time, they were cut transversely and the tissues were observed in a stereoscope. What were stained in deep red is considered viable, while those not, are considered unviable (**Figure 9**) [84].

The viability is 80% for *M. iltisiana*, this result agrees with what was found in [77] where they obtained 80% of viable seeds in *M. schiedeana*. While the seeds of *M. pugana* registered 67% [76]. Reference [75] reported that in species of the same genus from southern North America it has an average of 50% viability, similar results are found in *M. vallartensis* and *M. pacifica* with 50% and 53% respectively (**Figure 10**). These percentages are lower than those recorded in other *Magnolia* species. For example, in [74] found 100% viability in seeds of *M. dealbata* Zucc. [85]. In Ref. [80] obtained 92% for *M. perezfarrerae* A. Vázquez & Gómez-Dominguez [86] and 87.5% in *M. sharpii*.

2.6.3 Pre-germination treatments, germination tests, and dormancy types

Four treatments were used for *M. iltisiana* [81]. In *M. pugana* six treatments were carried out [76]. Three treatments were performed for *M. pacifica* and *M. vallartensis* [87]. The germination of *M. iltisiana*, *M. pacifica*, *M. pugana* and *M. vallartensis,* is of the epigeal type, the embryo developed a pair of foliaceous cotyledons, a flaccid fatty endosperm, a hypocotyl and a radicle, confined to the micropile area [76, 81, 87], (**Figure 11**).

In general, the percentage of germination obtained for *M. iltisiana, M. pacifica, M. pugana* and *M. vallartensis* are low. The highest germination recorded are 60%, 21%, 52% and 12%, respectively. In these studies, it is found that for *M. pacifica*, cold stratification treatment promoted the highest number of germinated seeds. Similar results have been obtained in [77] for *M. schiedeana*, with this same treatment with 84% germination. Another important finding is that the manual aryl removal treatment for *M. iltisiana, M. pugana,* and *M. vallartensis* are the most successful. These results coincide with those of reference [80] reported for *M. perezfarrerae* (64%) and *M. sharpii* (73%) with the mechanical scarification treatment (i.e., the seeds are placed in purified water and then the aril is manually

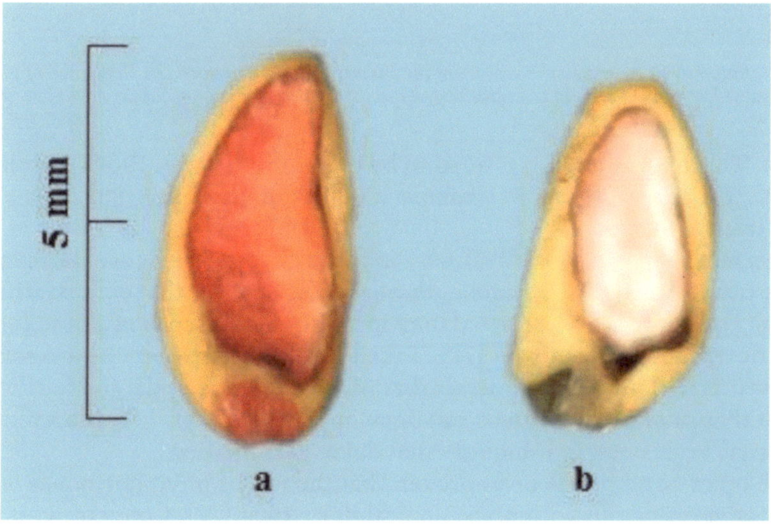

Figure 9.
Staining of seed tissues with tetrazolium in the viability test in Magnolia pugana*: a) viable seed, and b) non-viable seed.*

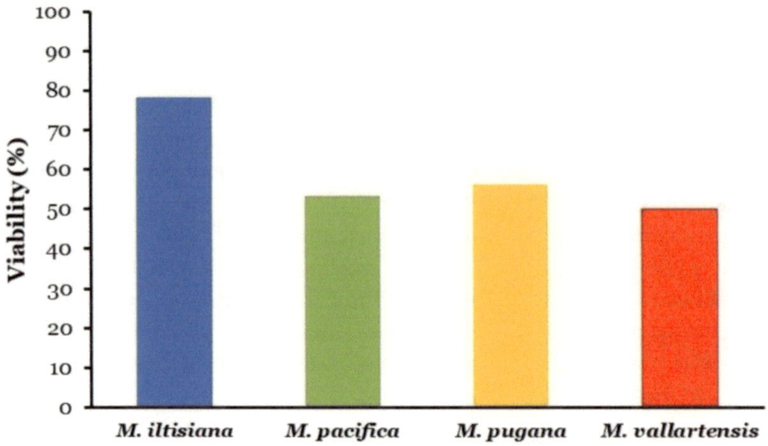

Figure 10.
Viability percentages for: Magnolia iltisiana, M. pacifica, M. pugana, *and* M. vallartensis.

Figure 11.
Germination process of M. pugana: *a) exposure of the endosperm and hypocotyl, b) elongation of the hypocotyl and cotyledon, c) leaf cotyledons and testa flushing, d) seedling with leaf primordia exposed above ground.*

removed). This treatment also proved to be effective with 90–100% germination in seeds of *M. dealbata* [74] and *M. champaca* (L.) Baill. ex Pierre [88] (73%) under the same treatment [89].

Low germination rates (< 70%) may indicate that the seeds are dormant and cannot be broken [90]. The results in these studies suggest that cold stratification treatments and manual aryl removal may indicate the presence of physiological and chemical dormancy, respectively [72]. Taken together, these results are consistent with reference [29] who recommends that *Magnolia* seeds undergo stratification periods so that immature embryos can develop, while in Ref. [91] reported that Magnoliaceae aryls contain inhibitors that delay germination.

On the other hand, it has been proven that the use of phytohormones is a promoter of germination with physiological dormancy [72]. Conversely, it was found that the phytohormone treatments used in the experiments on *M. iltisiana* and *M. pugana* did not increase the germination percentage, suggesting that the seeds of these species do not possess physiological dormancy. Although the seeds of

Species	Dormancy type	Reference
M. iltisiana	Chemical, physical and mechanical	[81]
M. pacifica	Chemical	[87]
M. pugana	Chemical	[76]
M. vallartensis	Chemical	[87]

Table 5.
Types of dormancy reported in Magnolia iltisiana, M. pacifica, M. pugana *and* M. vallartensis.

M. iltisiana have chemical dormancy, physical dormancy is also identified, that manifests itself through the development of a lignified testa that prevented the absorption of water, and the third type of dormancy is mechanical, which in this case the head exerted too much pressure on the embryo, delaying germination [81]. In the studies conducted for *M. pacifica*, *M. pugana* and *M. vallartensis* it was found that seeds do not have physical dormancy because they have the capacity to absorb water [76, 87] (**Table 5**).

3. Future scenarios

3.1 Vulnerability in climate change scenarios

Climate change is an important driver for future distributions of tree species, as it is expected to modify environmental conditions critical for plant populations maintenance. An analysis of 40 species of North American tree species from the temperate zone [92] revealed the higher vulnerability to climate change for the species growing in a colder climate, including the high-elevation mountain trees. In tropical mountain landscapes of Andes, the changes in habitat suitability are expected to be of a mixed character, while favorable for some tree species they may be prejudicial for others, with an overall general increase of species risk of extinction found in approximately 20% of tree species [93]. The increase of the risk of tree species extinction from climate change is related to the projected scale of the climate zones shift, which is highly variable across the globe [94]. In the situation of the terrain with complex irregular topography the extent of habitat suitable for species may be particularly sensitive to climate change [95], as the habitat tolerances of the mountain flora are generally narrow and the distribution shift upwards in the mountains frequently means the reduction of populations. At the same time, the complex topography may offer the high heterogeneity of habitat, which could provide opportunities for emerging local refugia, detectable at a fine spatial scale [96]. In the case of *Magnolia*, the narrow distribution ranges of species may constitute a risk factor, that could lead to the disappearance of the suitable habitat in some of the climatic projections for the end of the 21st century, as it was found in the case of the Andean species *M. mercedesiarum* D.A. Neill, A. Vázquez & F. Arroyo [97, 98].

An important step in the assessment of tree species vulnerability and extinction risk in the climate change scenarios is the analysis of habitat suitability dynamics. The habitat suitability belongs to the core concepts of the ecological niche theory, particularly of great importance in the field of plant ecology, as plants cannot evade adverse environmental conditions by sheltering or migrating within the single generation [99]. The assumption that healthy plant populations stay in the equilibrium with the contemporary or recent environmental conditions is central for accessing habitat suitability through niche modeling and projecting [100]. The climatic envelope models are a particular class of habitat suitability and niche models that uses the climatic variables as an approximation to the maximal extent of the habitat suitable to support species [101]. This class of models is widely used for the analysis of species-climate relations in the current climate, and for the heterochronic studies using past and future climate strata, as well as for predicting species distributions.

After accepting the equilibrium assumption, we used the field observations of the presence of three *Magnolia* species to develop the corresponding climatic envelope models, which captured the range of the environmental conditions that are suitable for species persistence in time and space. Further projections of the habitat suitability in the climate change scenarios in the next 80 years allowed us to detect the probable habitat suitability changes for *Magnolia* species, and to evaluate species vulnerability to climate change.

3.1.1 Methods

The species presence dataset of three taxa of the *Magnolia pacifica* complex [*M. pacifica* s.l. (including *M. talpana*), *M. pugana* s.l. (including *M. granbarrancae*), and *M. vallartensis*] was compiled from herbarium specimens at IBUG, IEB, ZEA, WIS, MEXU, XAL, MO, MICH, field observations cited in taxonomic literature [11, 13, 16], virtual images at Naturalista [102]. Monthly averaged climatic variables from WorldClim 2.1 dataset [103] available for the recent past (1970–2000) were taken as an approximation to current conditions, while the future conditions under two CO_2 emission scenarios were taken from the downscaled projections of the general circulation model CanESM5 [104]. The detailed description of the species distribution modeling procedure and predictions are available in Mendeley Data repository [66].

3.1.2 Results and discussion

As it was expected, the climatic envelope models recovered the suitable habitat extent larger than the known species distributions. The reasons of the overestimation are discussed in Shalisko *et al.* [66]. However, in the continental scale, the high suitability was predicted close to the occurrence records, the most distant grid cells identified as suitable were separated from known species occurrence sites in less than 150 km. The estimated suitable for the species presence zone in 1970–2000 varied from 9560 km^2 for *M. vallartensis* to 23940 km^2 for *M. pugana* s.l. (**Figure 12**). The suitable areas were well separated from another two species in the case of *M. pugana* s.l.,overlapped with *M. pacifica* s.l. In the case of the *M.6vallartensis* the estimated overlap of potentially suitable habitat with that of *M. pacifica* s.l. was of 4053 km^2, which equals to 42% of the entire high suitability zone for *M. vallartensis*.

Despite the systematic overestimation of the suitable area, the climatic envelope models are useful for the evaluation of the species vulnerability to climate change, as the same bias applies to the prediction of habitat suitability in current conditions and future projections. The changes in the area with suitable conditions may be proportional to the changes in true potential distribution.

The dynamics of suitable area in SSP2–4.5 scenario [105] (**Figure 13**) was favorable for *M. pugana* s.l., as the size of the potentially suitable area in 2080–2100 was about 13% larger than in current conditions, and almost all grid cells labeled as suitable in recent past persist at the end of 21st century. The models for both *M. pacifica* s.l. and *M. vallartensis* predicted suitable habitat reduction which is particularly fast in the second half of the century, with loss of about 45% of the suitable habitat in the case of the former species and 53% in the least. Interestingly, in the

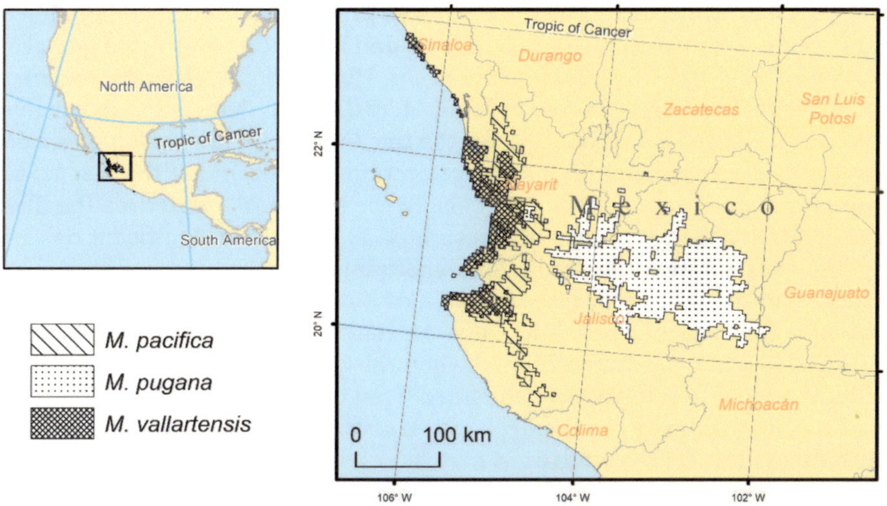

Figure 12.
Areas of habitat suitability higher than ESS threshold in at least 50% of cross-validation runs of the climatic envelope model for recent (1970–2000) climatic conditions. The key map on the left part shows the location of the right frame within North America.

current climatic conditions, the habitat suitability for *M. vallartensis* was found to be high not only in the locations close to the known distribution, but also in the separate coastal zone northwards, and the persistence of the habitat suitability in future scenarios was higher in the northern area, where species observations are unknown. In the case of *M. pacifica* s.l.the future high habitat suitability in SSP2–4.5 scenario was predicted roughly in the same geographic zones as was identified for current conditions.

The baseline SSP3–7.0 scenario [105] produced habitat suitability projections that are concerning in terms of species survival (**Figure 13**). In the case of three species, the fast decline in habitat suitability was predicted from the middle of the century, resulting in a loss for the end of the century of 66% of the suitable area in the case of *M. pugana* s.l., dramatic 92% loss of habitat suitable for *M. pacifica* s.l. and complete disappearance of the habitat of *M. vallartensis*.

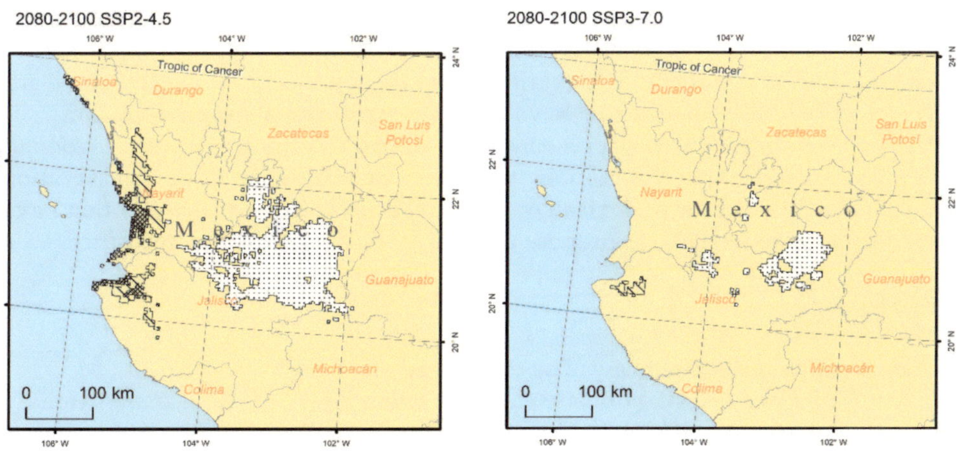

Figure 13.
*Areas of habitat suitability higher than ESS threshold in at least 50% of cross-validation runs of the climatic envelope model for projected future (2080–2100) climatic conditions under SSP2–4.5 and SSP3–7.0 scenarios. Symbology and extent is the same as in **Figure 12**.*

The uncertainty associated with habitat loss projections remained high due to the limitations of climatic envelope modeling and the uncertainty from global circulation models and CO_2 emission scenarios. However, the general trend of probable habitat loss for *M. pacifica* s.l.and *M. vallartensis* detected in the analysis was not compromised by accounting for the uncertainty, as the similar trend of change between current conditions and projections for the end of the 21st century was found in the 95% confidence intervals. In the case of *M. pugana* s.l. the likely habitat suitability change cannot be interpreted as the sign of significant risk when taking into consideration the prediction uncertainties.

Similarly to reference [95] interpretation of the effect of habitat suitability reduction in the species survival, we consider that the risk of species extinction from habitat loss may be overestimated when the data was analyzed in coarse-scale, as the local small size refugia were excluded from consideration. The true vulnerability of species to climate change depends on several factors outside of the scope of the current analysis. Tree populations may have a lag in reaction to climate change as the long-living sessile organisms [106]. The result of this lag could be the absence of the immediate disappearance risk for adult tree individuals, that could successfully tolerate significant environmental stress, but the reduction of the reproductive success required for the populations maintenance. In many species of the North American trees, the observed distribution is not entirely concordant with the current climate, as the long-living organisms may present the ‘extinction debts’ and ‘colonization credits’ at some parts of their actual or potential ranges [106]. In the case of *Magnolia* species, the lag between the loss of the climatic suitability and the actual population disappearance may be at least several decades required for the eventual death of adult representatives of the populations. Another aspect that was not accounted for in the vulnerability analysis is the possibility of the wide environmental tolerance within the populations that were not manifested in the current biotic and abiotic conditions, but that could contribute to species survival in climate change scenarios.

4. Conservation strategy

4.1 Conservation status

More than 350 *Magnolia* species have been described worldwide, and of a total of 314 that have been evaluated in their conservation status, 48% (147 sp.) have been classified as threatened with extinction in the wild [107]. However, there are still 34% of *Magnolia* species that have not been evaluated mainly due to data deficiencies, despite that *Magnolia* conservation has become a major international task [108]. Therefore, in this section, we present an evaluation of the conservation status of the three western Mexican *Magnolia* species (section *Magnolia*) that have been evaluated so far [18], to propose and implement conservation measures.

4.1.1 Methods

Conservation status assessment. The threatened status of *M. pacifica* s.s., *M. talpana*, *M. pugana* s.s., *M. granbarrancae*, *M. vallartensis,* and *M. iltisiana* were assessed here using the IUCN Red List Criteria (criteria B1ab + B2ab, [109]) and the GeoCAT cloud software tool [110]. The Extent of Occurrence (minimum convex polygon, EOO) and the Area of Occupancy (grid cell area with occupancy, AOO) of each taxon were delimited from georeferenced records obtained from IBUG herbarium specimens, the GBIF, Tropicos.org, REMIB-CONABIO, REBIOMEX databases, and field data. AOO was based on the IUCN default cell width of 2 km. The criterion C-2 (Genetics) of the Method of Extinction Risk Evaluation of Plants in Mexico (MER) from the Mexican Official Norm 059 [111] was used to contribute partially to the MER assessment. Criterion C-2 proposes that (1) if the population has heterozygosity (He) < 10–20% (depending on the molecular marker used) and (2) a genetic differentiation (Gst or Fst) > 20%, it has a higher threatened status or extinction risk.

4.1.2 Results and discussion

The estimated EOO (km^2) and AOO (km^2) were: for *M. pugana* s.s. =1,259.3, 96, respectively; *M. granbarrancae* = 0.7, 12; *M. pacifica* s.s. =1,216.2, 72; *M. talpana* = 91.8, 32; *M. vallartensis* = 124.0, 44; and *M. iltisiana* = 19,444.2,196 (**Table 6**). These western Mexican *Magnolia* species are in an endangered status because of their higher fragmented populations, lower genetic diversity, and narrower extent of occurrence when compared with other threatened *Magnolia* species [48].

M. granbarrancae and *M. talpana* have extremely narrow geographical ranges (EOO < 100 km^2) that are in accordance with the IUCN criteria for the category of Critically Endangered species: B1ab (iii, v). Populations of these species are small and severely fragmented (criterion B1a), and present a continuing decline in the area, quality of habitat [criteria B1b (iii)], and number of mature individuals [criteria B1b (v)] [109]. Moreover *M. granbarrancae* fulfill with the IUCN criterion C1 for Critically Endangered species, having a known total number of 74 mature individuals for the species, and an estimated or projected continuing decline of at least 25% in the next three years because the flooding of its habitat by the construction of the El Zapotillo damp on the Río Verde.

The genetic diversity of a species is an important indicator of its conservation status due to its positive correlation with the capacity to adapt and overcome abiotic and biotic changes. The genetic diversity of the three species studied here is considered lower than the average (H = 0.22) [45], and it is even lower than that of *M. sharpii* (I = 0.56), a species categorized as endangered because of its narrow EOO (2,228 km^2), severely fragmented and degraded habitats, and the fact that it is known from only five locations [48, 107]. *M. pugana* together with *M. granbarrancae* had a total heterozygosity H_T of 0.158, *M. pacifica* together with *M. talpana*, of 0.175, and *M. vallartensis* of 0.171, this low genetic diversity of the three species is in accordance with criterion C-2 (intrinsic biological vulnerability with genetic heterozygosity <20%) of the Mexican Standard NOM-059-SEMARNAT MER for being considered in the category of Endangered species [111].

The IUCN criteria do not consider the levels of genetic diversity and differentiation for assessing extinction risk, but *M. pugana* and *M. granbarrancae* suffer of very low genetic diversity [11] and very low number of individuals, so these two species should be categorized as Critically Endangered even when *M pugana* s.s. has an area of occupancy of 96 km^2. It is noteworthy that despite *M. pugana* s.s. EOO and AOO are in accordance with the IUCN Endangered category, their highly fragmented and isolated small populations suffering a seasonal dry and extreme climate, and its low genetic diversity make consider this species as Critically Endangered. *M. pugana* s.l. (including *M. granbarrancae*) has even lower genetic diversity and higher genetic differentiation than the other western Mexican *Magnolia* species studied. This fact, together with the fact of having more fragmented, more isolated, smaller populations, and being surrounded by a seasonally drier environment [103], make *M. pugana s.s.* and *M. granbarrancae* be proposed to be included in the category of Critically Endangered of extinction, as it had been previously cataloged by [112] for *M. pugana* s.l.

	Magnolia iltisisana	*Magnolia granbarrancae*	*Magnolia pacifica*	*Magnolia pugana*	*Magnolia talpana*	*Magnolia vallartensis*
IUCN Red List category [107]	Vulnerable	—	Endangered	Endangered	—	Critically Endangered
Current threats	Anthropogenic factors, seed predation by squirrels, rapid degradation of fallen indehisced fruits, shade intolerance, low seed dispersal, livestock, illegal logging	Agricultural expansion, construction of the El Zapotillo dam that will flood a large part of the species' population, climate warming	Illegal logging, livestock, mining, agricultural expansion, avocado and coffee plantations, fruit orchards (guava, citrus), forest fires	Isolated populations, fragmented habitats, high deforestation rate, low regeneration, changes in land use, forest fires	Agricultural expansion, illegal logging, avocado and coffee plantations, fruit orchards (guava, citrus), forest fires	Restricted distribution, highway construction for Puerto Vallarta bypass, forest fires, insect pests, global warming, agricultural expansion
No. of individuals	43 ind. in 20 0.1 ha plots, Villas de Cacoma	Only 74 known ind. From RV	48 ind. in six 0.1 ha plots, Cerro La Bufa +72 ind. in San Sebastián del Oeste region	735 known ind. From all known populations ASL, ALV, APV and others	40 ind. in 20 0.1 ha plots +54 outside of the plots, Talpa river watershed, maple forest	27 ind. in two 0.1 ha plots, from APM, 187 known ind. From APM, LL, PV and others
Ex-situ records	2	0	55	209	0	19
Genetic diversity	Unknown	Very low	Low	Very low	Low	Low
Seed productivity	High	Middle	Middle	High	Low	Low
Known populations in 1994	3	0	7	6	0	0
No. populations in 2020	23	3	14	14	5	6
Radius (km)	45	1.5	50	30	15	16
Biogeographic provinces	1	2	3	2	1	2
Altitude (m a.s.l.)	800–2400	1000–1240	750–2250 m	1300–1800	1050–1800	100–1100

	Magnolia iltisisana	*Magnolia granbarrancae*	*Magnolia pacifica*	*Magnolia pugana*	*Magnolia talpana*	*Magnolia vallartensis*
Latitudinal	19°10′–19°50′	20°45′–20°43′	20°30′–21°38′	20°46′–21°15′	20°12′–20°18′	20°20′–20°35′
EOO (km^2)	19,444.2	0.7	1,216.2	1259.3	91.8	124.0
AOO (km^2)	196	12	72	96	32	44
Updated category from IUCN criteria: B1ab (iii, v), B2ab (iii,v), or C1	Vulnerable B1b (iii,v) B2b (iii,v)	Critically Endangered B1ab(iii,v), C1 Low genetic diversity	Endangered B1ab (iii,v) B2ab (iii,v)	Critically Endangered B1ab (iii,v) B2ab (iii,v) Low genetic diversity	Critically Endangered B1ab(iii,v)	Critically Endangered B1ab(iii,v)

Table 6.
Relevant conservation data to guide establishing conservation priorities for Magnolia *section* Magnolia *in western México. EOO: Extent of occurrence, AOO: Area of occupancy.*

Magnolia vallartensis should also be considered as Critically Endangered, as it has been categorized by [107], due to having an EOO of only 124 km^2, low number of individuals, high deforestation rates, forest plagues and fires, fragmentation, climate change, cattle raising, roads and urban growth. Both, the EOO and the AOO of *M. pacifica* s.s. are in accordance with the IUCN category of Endangered (EOO < 5,000 km^2 and AOO <500 km^2). Furthermore, in the case of *M. pacifica* s.l. (including *M. talpana*) and *M. vallartensis* the extent of occurrence is projected to suffer a severe reduction in the next 80 years as a consequence of the shrinkage of the areas of high habitat suitability, estimated under the highly probable SSP3–7.0 and less probable SSP2–4.5 scenarios. The scope of the projected extent of occurrence reduction in the case of *M. vallartensis* may bear this species into the critically endangered status by the end of the century [criteria B1b (i, iii)]. In the case of *M. pacifica* s.l. the projected reduction of the area of suitable habitat for the end of the century is highly significant, however, may be not enough to move species to the critically endangered category by using only this criterion.

Despite *M. iltisiana* has an AOO <500 km^2 (196), and that it has some populations with severe fragmentation, as those near Morelia city (in Michoacán), several populations of this species do not have severe fragmentation and inhabit in the protected area of Sierra de Manantlán Biosphere Reserve. Therefore *M. iltisiana* should maintain its category of Vulnerable, as it has been suggested by [107].

Magnolia granbarrancae which is located to the northeast side of the Santiago river canyon (composed by the RV locality), *M. pugana* s.s. (composed by ALV, ASL and APV), and the populations of *M. pacifica* s.s., *M. talpana* and *M. vallartensis* should be considered as five separate units of conservation. The implementation and enforcement of *in situ* and *ex situ* conservation actions should protect and preserve at least one locality of each population of *M. pugana* s.s., and *M. granbarrancae*, as well as the most divergent localities of *M. pacifica* s.s. (CSJ and SS), *M. talpana* (BA) and *M. vallartensis* (PV and APM). The BA locality of *M. talpana* ("Bosque de Arce", maple forest) conserve the highest genetic diversity, which coincides with being one of the most important forests in terms of plant species richness, endemism, and floristic composition for western Mexico [113, 114]. Greater efforts must be made to preserve all of these taxa, and a higher focus is required to protect *M. pugana* s.s and *M. granbarrancae*, characterized by low levels of genetic variation and highly fragmented and small populations. All these *Magnolia* species must be fully evaluated with the MER method for inclusion in the list of endangered species of the official Mexican Standard NOM-059. Education, conservation, management, and ecological restoration plans are badly needed to decrease their threatened status and raise awareness of the fate of these important species in extinction risk.

4.2 Propagation and reintroduction

Magnolias, like other plant species, may be reproduced both sexually (by seeds) and asexually (through plant tissues).

4.2.1 Sexual reproduction

This kind of propagation involves genetic recombination, which provides a genetic variability that improves the plant's ability to adapt to its environment [29, 115], ensuring that long-term survival by reducing the risk of suffering a bottleneck effect, which puts a species in danger of extinction [116]; Also, more vigorous seedlings are generated and the propagation is easier and cheaper than asexual reproduction, on the other hand, the plants take longer to reach maturity and bloom. In section "2.6 Pre-germination and seed dormancy treatments" of this chapter, the aspect of sexual reproduction in magnolias of Western Mexico is addressed more extensively.

4.2.2 Asexual propagation

Asexual reproduction has been only reported for *M. tamaulipana* of sect. *Magnolia* and for species of sect. *Macrophylla*. Asexual propagation occurs through stems, shoots and roots and is more convenient for horticultural purposes, where it is sought to preserve certain characteristics through generation of clones and flower faster than those propagated by seeds [29], however, the methods used are more expensive and transportation is more complicated [117]. Although no studies have been conducted on vegetative propagation in the Mexican magnolias of the sect. *Magnolia*, we list the most used horticultural methods for the *Magnolia* genus [29, 117]:

4.2.2.1 Cuttings

This approach has the advantage of reducing costs in large-scale production and the plants obtained are more uniform, but to ensure survival it is necessary to have better-equipped facilities and develop specific protocols for Neotropical magnolias. Method for cuttings in *Magnolia* could be found in the following references: [29, 117, 118].

4.2.2.2 Layering

It is recommended when you want to get only a few plants. Produces larger plants in a shorter period of time, but requires more space and labor [29]. More information about this technique can be avaible in the following references [29, 117, 118].

4.2.2.3 Harnpariphan's technique.

Here named as such, is an intermediate cutting-layering technique. In the cutting a reed incision is made 5 cm above the basal cut, a small piece of plastic is inserted between the reed and the stem to ensure an acute angle separation. A paste with fungicide and rooting hormone (1:1) is applied to the exposed sides of the incision. Subsequently, the wound area is covered with a substrate, securing it as layering with a piece of plastic tied at the ends with ropes. A small segment is cut from the basal part of the cutting-layer in water, to ensure that the air does not interrupt the flow of water through the conductive vessels. The basal part must be in contact with the water and the covered region of the wound must remain in the air, taking care to always keep the layering moist, making sure that the plastic has small holes that allow the excess water to drain. Leaf areas with lesions or infections should be trimmed and the water always kept fresh by making replacements when necessary [117].

4.2.2.4 Grafting

It is an easy and cheap method but requires more space, labor and time. Recommended for species and cultivars difficult to root [29], the graft should be done between genetically closest species since there is better long-term compatibility [119]. It is a great alternative for threatened species with low fertility [117]. For major types of grafts of magnolias see the following references: chip-budding [29, 118], side-grafting [29], wedge or cleft grafting [117], crown grafting [117], canutillo graft [117], shield budding [117], patch budding [117], and approach grafting [117].

4.2.2.5 Tissue culture

It involves the regeneration of a whole plant from a small portion taken from any part of a parent plant, producing enormous quantities of plants from a few cells. This method, in particular, represents a difficult and expensive task, coupled with the fact that some species may be more difficult to propagate by this means, a situation that is attributed to the high content of phenolic acid in magnolias, which can inhibit the growth of the crop, requiring frequently change the explant to a fresh medium, it has also been observed that magnolias tend to generate vitrified growth under tissue culture, which does not generate roots and rarely reverts to normal growth [29]. Phenolic acid content is lower in magnolias after dormancy is broken, so it is better to use dormant branches to obtain explants [120]. More information about this technique can be found in Refs. [29, 120, 121].

4.2.3 Reintroduction

The goal of reintroduction is to establish a viable population of any species in the wild and is essential to increase its long-term survival and to reestablish key species in an ecosystem and restore its natural biodiversity, so it must be carried out within the area of distribution and primitive natural habitat of the target species [122]. For a reintroduction to be successful, it should be considered the awareness of the population and community participation, the planting time and composition of the individuals in quantity and quality (**Figures 14** and **15**). It must be ensured that the site has the appropriate biotic and abiotic requirements for the species in all its life stages, considering seasonal and post-establishment needs, continuous monitoring and management is required to provide feedback [123]. In the case of magnolias, it has been found that they belong to an intermediate and late-successional state, so reintroduction and reforestation projects must consider planting individuals under a pre-existing plant cover [124]. In tropical magnolias, no tolerance to prolonged dry seasons has been found, so to plant them, humid regions, well-drained sites with slightly acidic soils (pH close to 6) and little compacted should be chosen [117]. For transplantation to the ground, reference [117] recommends loosening the soil first, taking care not to injure any roots when removing them from the bag, always keeping the soil moist, adding mulch around the stem to keep moisture and avoid weeds, and, in case of if necessary, apply dilute phosphoric acid to lower the soil pH.

Figure 14.
Course-workshop on conservation challenges for Magnolia *conservation in the subdeciduous forest of the Nahuan community of Ayotitlán, in the sierra de Manantlán, western México. Photos by A. Vázquez-García.*

Figure 15.
Left: Reintroduction of Magnolia pugana *by Dra. Rosa Romo and her assistants near arroyo san Nicolas. Central: Plantation of* Magnolia ofeliae *at the campus of UDG-CUCBA, Zapopan, by the kids: Alberto Aguilar and Quetzalcóatl Vázquez. Right: Urban plantation of* Magnolia mexicana *at Jardínes Alcalde, Guadalajara by Jesús Cortés and Paula Cortés. Photos by M. Muñiz-Castro and A. Vázquez-García.*

4.3 In-situ and ex-situ conservation

4.3.1 In-situ conservation proposals

Of the six species of *Magnolia* sect. *Magnolia* five of them have populations within Protected Natural Areas; *M. iltisiana* is located within the Sierra de Manantlán Biosphere Reserve, a federally protected area, while *M. talpana* is located within the Bosque de Arce State Park and *M. granbarrancae* in the Natural Formation of State Interest Barrancas de los Ríos Santiago y Verde, *M. pacifica* has one of its populations in the Sierra de San Juan Ecological Reserve in Nayarit and in the Cuenca Alimentadora del Distrito Nacional de Riego 043 (APRN-CADNR-043), and *M. pugana* has populations within the La Primavera Flora and Fauna Protection Area and in the Natural Resources Protection Area of Cuenca Alimentadora del Distrito Nacional de Riego 043 (APRN-CADNR-043, southern Zacatecas polygons). However, for some of these species it is necessary to establish additional measures for their protection [13].

In the case of *M. pacifica*, it is important to create a protection zone in its type locality, in San Sebastián del Oeste, Jalisco, where it is currently threatened and its population has been reduced by deforestation, agricultural expansion and mining. For *M. pugana*, it is proposed to create a protection zone in the perimeter of Arroyo La Virgen, which is located near Rancho San Nicolás in Zapopan, Jalisco. This stream has relict elements of cloud forest, so declaring it a Protection Area would benefit the conservation of other species such as *Populus luziarum* A. Vázquez, Muñiz-Castro & Padilla-Lepe [125] (CR), an endemic species found only in two Zapopan ravines.

With *M. vallartensis*, it is recommended to decree a protection zone in the Palo María stream basin, since it is one of its main distribution sites in addition to having other endemic species such as *Pinus vallartensis* Pérez de la Rosa & Gernandt [126] and *Miconia vallartensis* Zabalgoitia, Figueroa & Muñiz-Castro [127], which would contribute significantly to the conservation of Puerto Vallarta's biodiversity. In this particular case, the importance of establishing a protection zone and rescuing this species also rests on the fact that this species was declared an emblematic tree of the Municipality from which it bears its name, Puerto Vallarta.

Although all these species, except for the new ones described here, are under some risk category of the IUCN Red List, at the national level, only one of them, *M. iltisiana*, is found in the Official Mexican Norm NOM-059 on Environmental Protection for native species of flora and fauna; low genetic diversity of

M. granbarrancae (Rio Verde locality), *M. pacifica* s.s., *M. talpana, M. pugana* s.s., and *M. vallartensis*, agrees with the C-2 criterion of NOM-059 MER [18], therefore that it would be appropriate to integrate them into this Norm as Endangered species.

It is vitally important to create integrated management and conservation strategies according to each species, which include reaching stakeholders of the communities where these magnolias are distributed since the first step is to increase conservation awareness and foster appropriation of their natural resources. An example is a Workshop held by some of the authors of this chapter in the Nahua community of Ayotitlán, where *M. jaliscana* is located, in which they taught about the importance of conserving magnolias and of its propagation, as well as the importance of ecosystem services (**Figure 14**).

Alternatively, Wildlife Conservation Management Units (UMA for its acronym in Spanish) can be created in which, in addition to conserving the site, the owner obtains benefits through the sustainable use of its natural resources and can be beneficiaries of subsidies for the conservation and sustainable use of wildlife native at UMA [128].

Property owners can also be creditors of the Payment for Environmental Services (PES), which is a program whose purpose is to promote the recognition of the value of the services provided by ecosystems by creating a market for them [129].

4.3.2 Ex-situ conservation

The ex-situ conservation centers aim to reduce the risk of extinction of threatened species and act as a complement to in-situ conservation by supporting wild populations with the reintroduction of specimens and restoration of habitats, acting as gene banks, promoting research and continuing to raise social awareness as elements of diffusion and environmental education [130]. For *Magnolia*, there are 9,918 ex-situ records of 152 Magnoliaceae species, however less than half of the most threatened taxa are represented [107], Latin America and the Caribbean region are of particular concern because many of their endemic Magnoliaceae are

Species	University of Guadalajara, Zapopan	Vallarta Botanic Garden	National Center of Genetic Resources, Tepatitlán
M. dealbata	0	0	1
M. grandiflora	11	1	1
M. iltisiana	0	2	0
M. jaliscana	1	0	0
M. mexicana	0	0	1
M. oaxacensis	0	3	0
M. ofeliae	2	4	0
M. pacifica	17	38	0
M. pugana	180	29	0
M. rzedowskiana	0	1	0
M. tarahumara	80	35	0
M. vallartensis	14	5	0

Table 7.
Magnolia ex-situ collections from Jalisco, Mexico. M. mexicana *DC. [132],* M. oaxacensis *A. Vázquez [14],* M. rzedowskiana *A. Vázquez, Domínguez-Yescas & Pedraza-Ruiz [133],* M. tarahumara *(A. Vázquez) A. Vázquez [15].*

not represented in ex-situ collection [131]. In Jalisco, some of the *Magnolia* ex-situ collections for 2020 are here presented (**Table 7**), and particularly the species of sect. *Magnolia* discussed in this chapter are cultivated.

In March 2020, a Magnolia Conservation and Propagation Workshop was held in the Nahua community of Ayotitlán, in which a small greenhouse was installed so that the community could reproduce *M. jaliscana*, currently, they have 20 seedlings (**Figure 14**).

4.3.2.1 Plantations of M. pugana

Dr. Rosa de Lourdes Romo Campos has dedicated herself for several years to the conservation of *M. pugana*, which has led to a series of reintroduction plantations for this species, whose origin is from San Lorenzo, Zapopan. These plantations can be located in the map (**Figure 1**) with a tree symbol. In 2013 two plantations were carried out, one in the "Los Colomos" urban forest and another near the "Balneario El Encanto" in Zapopan, Jalisco, both with 50 plants; in 2016 one plantation with 30 individuals were made near highway San Cristobal de la Barranca-Tesistán; in 2018 one plantation with 25 trees was carried out in "Los Colomos" urban forest and another with 200 individuals in San Esteban, Zapopan, Jalisco was executed; in 2020 only one plantation was made in Parques de Tesistán, Zapopan, with 18 trees.

5. Conclusions

1. The *Magnolia pacifica* species complex shows high morphological divergence and displays a remarkable pattern of allopatric radiation, where the species are distributed among five biogeographic provinces and or separated by canyons, along an elevational, latitudinal and continentality or moisture gradient. This is demonstrated by the two newly described taxa, including: *M. granbarrancae*, confined to the 500 m deep and narrow Rio Verde canyon, in the vicinity of Guadalajara city and *M. talpana* confined to the southern portion of the Talpa de Allende municipality, Jalisco, in the high watersheds of the ríos Camacho, Desmoronado and Talpa. Further fieldwork is needed in northern Nayarit, Sinaloa, Durango, Sonora and Chihuahua to enhance our understanding of the morphological divergence of this species complex.

2. The genetic structure of the *Magnolia pacifica* species complex exhibits two main groups separated along a moisture gradient by the Transmexican Volcanic Belt province and by distance. The western more maritime group differentiated into two genetic subgroups *M. pacifica* s.s. and *M. vallartensis*, and the eastern most continental group differentiated into two genetic subgroups *M. pugana* s.s. and *M. granbarrancae*. *M. pugana* s.l. is subject to a drier and more extreme climate, therefore having more deforested, fragmented, and isolated habitats, which leads to lower genetic diversity and a higher genetic differentiation. This differentiation and the physiographic barrier of the Santiago river canyon have structured *M. pugana* s.l in the two subgroups, *M. pugana* s.s. and *M. granbarrancae*. The *M. pacifica* s.s-*M. vallartensis* group exhibits genetic segregation in two subgroups, *M. vallartensis* and *M. pacifica*, separated by several canyon river barriers, but maintaining a partial gene flow at their southern ranges. The three species of the *M. pacifica* complex have lower genetic diversity than eastern Mexican *Magnolia* species which are considered as endangered. All main genetic groups and subgroups defined in this study should be considered as separate conservation units, and concerted efforts are needed to protect them.

3. Similar to other basal angiosperm lineages, beetle pollination has been associated with the Magnoliaceae, however, we have documented that hymenopterans could also play a significant role as pollinators in some species as demonstrated in *Magnolia vallartensis,* being pollinated by meliponid bees. The pollination efficiency between Hymenoptera and Coleoptera needs to be assessed.

4. Flower showed an incipient asynchrony, insufficient to prevent gene flow, while fruiting had a marked asynchrony among the three species, suggesting specialization by reducing competitors and enhancing dispersal. Flowering seasonality was significant for *M. pacifica* and *M. pugana* while fruiting seasonality was significant only for *M. pugana*. In terms of productivity *M. pugana* had a significantly higher productivity than *M. vallartensis* or *M. pacifica*, despite this large productivity, there is little or no recruitment of juveniles.

5. Floral scents in the three studied species of *Magnolia pacifica* complex display distinctive floral scent profiles, the chemical differences may have arisen from interaction with pollinators or the environment. The Jaccard similarity index between *M. pacifica - M. vallartensis* was 54.6%, while with *M. pugana* these showed 19.5% and 18.6% of similarity respectively. It is necessary to continue with studies focused on establishing the relationships between the components and the chemotaxonomic field including more *Magnolia* species populations to conform a robust database.

6. The conservation status of the western Mexican *Magnolia* species are currently as endangered status because of their highly fragmented populations, low genetic diversity, and narrower extent of occurrence compared with other threatened *Magnolia* species. But it is noteworthy that *M. pugana* s.l. group has even lower genetic diversity and higher genetic differentiation than the other western Mexican *Magnolia* species. This fact, together with the fact of having more fragmented, more isolated, smaller populations, and being surrounded by a seasonally drier environment, make *M. pugana* a candidate species to be included in the category of Critically Endangered, as it had been previously cataloged.

7. Climate change scenarios projected in the next 80 years suggest high vulnerability for two species (*M. pacifica* and *M. valartensis*). The area of high habitat suitability is expected to abruptly reduce for both species, which may result in persistence of populations only within the small local refugia, which are located within the extent of the current high suitability area. The species in-situ preservation effort may be particularly important in the areas that were found to have high habitat suitability in the climate change scenarios.

8. Given that there is still little knowledge on insect floral visitors and the floral biology of most *Magnolia* species, further studies including more field observations are needed. Studies of seed predation and dispersal are also highly recommended to understand the apparent lack of recruitment of juveniles.

9. The results of the percentages of viability differ from those of germination in *M. iltisiana, M. pacifica, M. pugana* and *M. vallartensis*. Therefore, it is necessary to continue with the studies of seed ecology, to determine whether there are other types of dormancy as morphological or morpho-physiological, as has been detected in other species of *Magnolia*. Considering that the highest percentage of germination is achieved with the manual removal of the aril, for its low cost and in greenhouse conditions, it can be a successful practice for the

propagation of seedlings of these species, which help reforestation and restoration of their populations in their natural environment.

Acknowledgements

We dedicate this chapter to the distinguished Professor of Natural Sciences Dr. Enrique Estrada Faudón, from University of Guadalajara, the western Mexican "Humboldt" for leading since 1985, the first large-scale propagation of *Magnolia pugana* to prevent extinction. To the University of Guadalajara (UDG), Centro Universitario de Ciencias Biológicas y Agropecuarias, Botanic Gardens Conservation International and Fondation Franklinia. Especially thanks to Dr. Adriana Natividad Avedaño López, Sergio Misael Gallegos Mendoza and Weerakit Harnphariphan for their support and advice in conducting this research. This work was supported by Secretaría de Educación Pública (SEP-México) through its PRODEP program PROMEP-NPTC/103.5/12/3418, through its PRO-SNI program and the P3E 2013-2018 project of CUCBA (UDG). We are grateful to Bob Price, Neil Gerlowski, Ralph Osborne†, from Vallarta Botanical Gardens, members of the Magnolia Society International, Global Conservation Consortium Magnolia (GCCM) and Toribio Quintero Moro from Ayuntamiento de Talpa de Allende for financial and or logistical support. To Patricia Castro-Félix, Anne Santerre, Ahtziri Socorro Carranza Aranda, Mario Ruiz López, J. Luis Navarrete Heredia, Alma Rosa Villlobos for their research collaboration. To supporters in field explorations: Abel Puente Casillas, Alejandro Zabalgoitia, Ana Teresa Nuño Rubio, Heriberto Justo Nieves, Jesús Padilla-Lepe, Jesús Truijillo, Miguel Cházaro, Ricardo Díaz Borioli, Rosa Murguía Araiza and Saúl Cristobal Gabriel.

Author details

José Antonio Vázquez-García[1*], Miguel Angel Muñiz-Castro[1], Alex Dahua-Machoa[1], Edison Antonio Osorio-Muñoz[2], Gerardo Hernández-Vera[1], Alondra Salomé Ortega-Peña[1], Rosa de Lourdes Romo-Campos[1], César Jacobo-Pereira[1], Noelia Álvarez de Román[3] and Viacheslav Shalisko[1]

1 University Center for Biological and Agricultural Sciences, of the University of Guadalajara, Zapopan, México

2 Research and Development Group in Sciences Applied to Biological Resources, Universidad Politécnica Salesiana, Quito, Ecuador

3 Botanic Gardens Conservation International, Richmond, United Kingdom

*Address all correspondence to: talaumaofeliae@gmail.com

References

[1] Vázquez-García JA, Neill D, Asanza M, Pérez ÁJ, Arroyo F, Dahua A, et al. Magnolias de Ecuador: En Riesgo de Extinción. Universidad Estatal Amazónica; 2016. 143 p.

[2] Domínguez-Yescas R, Vázquez-García JA. Flower of the heart, *Magnolia yajlachhi* (subsect. *Talauma*, Magnoliaceae), a new species of ceremonial, medicinal, conservation and nurse tree relevance in the Zapotec culture, Sierra Norte de Oaxaca, Mexico. Phytotaxa. 2019;393(1):21–34. doi: 10.11646/phytotaxa.393.1.2

[3] Jussieu AL. Genera plantarum. 1789. 281 p.

[4] Linnaeus C. Species plantarum. Stockholm: Salvius; 1753. 560 p.

[5] Wang Y, Liu B, Nie Z, Chen H, Chen F, Figlar RB, et al. Major clades and a revised classification of *Magnolia* and Magnoliaceae based on whole plastid genome sequences via genome skimming. J Syst Evol. 2020;58(5): 673–95. doi: 10.1111/jse.12588q

[6] Nooteboom HP. Magnoliaceae. In: Kubitzki K, Rohwer JG, Bittrich V, editors. Flowering Plants Dicotyledons. The Families and Genera of Vascular Plants, vol 2. Berlín: Springer; 1993. p. 391–401.

[7] Hebda RJ, Irving E. On the origin and distribution of magnolias: Tectonics, DNA and climate change. GMS. 2004; 145:43–57.

[8] Figlar RB, Nooteboom HP. Notes on Magnoliaceae IV. Blumea-Biodiversity, Evol Biogeogr Plants. 2004;49(1):87–100.

[9] García-Morales LJ, Iamonico D, García-Jimenez J. Nomenclatural remarks on *Magnolia* sect. *Macrophylla* (Magnoliaceae), with description of a new species from North America (Tamaulipas, Mexico). Phytotaxa. 2017; 309(3):238–44. doi: 10.11646/phytotaxa.309.3.4

[10] Baillon HE. Mémoire sur la famille des Magnoliacées. Adansonia. 1866;7:1–16,65–69.

[11] Vázquez-García JA. *Magnolia* (Magnoliaceae) in Mexico and Central America: a synopsis. Brittonia. 1994;46: 1–23. doi: 10.2307/2807454

[12] Vázquez-García JA, Pérez-Farrera MÁ, Goméz-Domínguez H, Muñiz-Castro MÁ, Sahagún-Godínez E. *Magnolia montebelloensis*, a new species in section *Magnolia* from Lagunas de Montebello National Park, Chiapas, México, with a key to Magnoliaceae of Chiapas. Phytotaxa. 2017;328(2):101–14. doi: 10.11646/phytotaxa.328.2.1

[13] Muñiz-Castro MÁ, De Castro-Arce E, Iltis HH, Guzmán-Mejía R, Cuevas-Guzmán R, Murguía-Araiza R, *et al.* Magnolias de Jalisco [Internet]. Guadalajara: Universidad de Guadalajara; 2019. 48 p. Available from: https://www.magnoliamexico2019.org/node/6

[14] Vázquez-García JA, Muñiz-Castro MÁ, De Castro-Arce E, Murguía Araiza R, Nuño Rubio AT, Cházaro-Basañez M de J. Twenty new neotropical tree species of *Magnolia* (Magnoliaceae). In: In: Salcedo-Pérez E, Hernández-Álvarez E, Vázquez-García JA, Escoto-García T, Díaz-Echavarría N, editors. Recursos Forestales en el Occidente de México: Diversidad, Manejo, Producción, Aprovechamiento y Conservación Tomo I. México: Universidad de Guadalajara; 2012. p. 91–130.

[15] Vázquez-García JA, Muñiz-Castro MÁ, Arroyo F, Pérez AJ, Serna M, Cuevas Guzmán R, et al. Novelties in Neotropical Magnolia and an addendum proposal to the IUCN Red List of

Magnoliaceae. In: Salcedo-Pérez E, Hernández-Álvarez E, Vázquez-García JA, Escoto-García T, Díaz-Echavarría N, editors. Recursos Forestales en el Occidente de México: Diversidad, Manejo, Producción, Aprovechamiento y Conservación Tomo II. México: Universidad de Guadalajara; 2013. p. 461–96.

[16] Vázquez-García JA, Carvajal S, Hernández L. L. *Magnolia pugana* (Magnoliaceae): A new combination in the complex *M. pacifica*. Novon. 2002; 12(1):137–141. doi: 10.2307/3393253

[17] Morrone JJ, Escalante T, Rodriguez-Tapia G. Mexican biogeographic provinces: Map and shapefiles. Zootaxa.2017;4277(2):277–9. doi: 10.11646/zootaxa.4277.2.8

[18] Muñiz-Castro MÁ, Castro-Félix P, Carranza-Aranda AS, Vázquez-García JA, Santerre A. Population genetics, species boundaries, and conservation in the *Magnolia pacifica* species complex along a continentality and moisture gradient in western Mexico. Bot Sci. 2020;98(3):500–516. doi: 10.17129/botsci.2551

[19] Osorio Muñoz EA. Identificación química y perspectiva medicinal de los aceites esenciales de hojas, semillas y flores de *Magnolia pugana*. Universidad de Guadalajara; 2020.

[20] Dahua-Machoa NA. Temporalidad de fenofases y micropropagación in vitro de tres especies relictuales de *Magnolia* del Occidente de México: implicaciones para su conservación in situ y ex situ. Univesidad de Guadalajara; 2018.

[21] Crepet WL, Friis EM. The evolution of insect pollination in angiosperms. In: Friis EM, Chaloner WG, Crane PR, editors. The origins of angiosperms and their biological consequences. England: Cambridge University Press; 1987. p. 181–201.

[22] Bernhardt P. Convergent evolution and adaptive radiation of beetle-pollinated angiosperms. Plant Syst Evol. 2000;222:293–320. doi: 10.1007/BF00984108

[23] Thien LB, Kawano S, Azuma H, Latimer S, Devall MS, Rosso S, et al. The floral biology of the Magnoliaceae. In: Hunt DR, editor. *Magnolias* and their allies. Egham, Surrey, U. K.: Royal Holloway, University of London; 1996. p. 37–58.

[24] Burmeister H. Handbuch der Entomologie, vol. 4, part 1. TCF Enslin, Berlin. 1844;586.

[25] Cresson ET. Descriptions of New Species of North American Bees. Proc Acad Nat Sci Philadelphia. 1878 Sep;30: 181–221.

[26] Kevan PG, Viana BF. The global decline of pollination services. Biodiversity. 2003;4(4):3–8. doi: 10.1080/14888386.2003.9712703

[27] Potts SG, Biesmeijer JC, Kremen C, Neumann P, Schweiger O, KuninWE. Global pollinator declines: trends, impacts and drivers. Trends Ecol Evol. 2010;25(6):345–53. doi: 10.1016/j.tree.2010.01.007

[28] Vásquez-Morales SG, Sánchez-Velázquez LR, Pineda-López R. Las semillas de la magnolia. La ciencia y el hombre. 2010;23(3):37–40.

[29] Callaway DJ. The world of magnolias. Portland, USA: Timber Press; 1994. 260 p.

[30] Hamrick JL, Godt MJW. Allozyme Diversity in Plant Species. In: Plant Populaiton Genetics, Breeding, and Genetic Resources. 1990.

[31] Veltjen E, Asselman P, Hernández Rodríguez M, Palmarola Bejerano A, Testé Lozano E, González Torres LR, et al. Genetic patterns in Neotropical

Magnolias (Magnoliaceae) using *de novo* developed microsatellite markers. Heredity. 2019;122:485–500. doi: 10.1038/s41437-018-0151-5

[32] Sosa V, Ruiz-Sanchez E, Rodriguez-Gomez FC. Hidden phylogeographic complexity in the Sierra Madre Oriental: The case of the Mexican tulip poppy *Hunnemannia fumariifolia* (Papaveraceae). J Biogeogr. 2009;36(1): 18–27. doi: 10.1111/j.1365-2699.2008.01957.x

[33] Ruiz-Sanchez E, Rodriguez-Gomez F, Sosa V. Refugia and geographic barriers of populations of the desert poppy, *Hunnemannia fumariifolia* (Papaveraceae). Org Divers Evol. 2012; 12:133–43. doi: 10.1007/s13127-012-0089-z

[34] Ruiz-Sanchez E, Specht CD. Influence of the geological history of the trans-mexican volcanic belt on the diversification of *Nolina parviflora* (Asparagaceae: Nolinoideae). J Biogeogr. 2013;40(7):1336–47. doi: 10.1111/jbi.12073

[35] Ruiz-Sanchez E, Ornelas JF. Phylogeography of *Liquidambar styraciflua* (Altingiaceae) in Mesoamerica: Survivors of a Neogene widespread temperate forest (or cloud forest) in North America? Ecol Evol. 2014;4(4):311–28. doi: 10.1002/ece3.938

[36] Bryson RW, García-Vázquez UO, Riddle BR. Phylogeography of Middle American gophersnakes: Mixed responses to biogeographical barriers across the Mexican Transition Zone. J Biogeogr. 2011;38(8):1570–84. doi: 10.1111/j.1365-2699.2011.02508.x

[37] Bryson RW, Riddle BR. Tracing the origins of widespread highland species: A case of Neogene diversification across the Mexican sierras in an endemic lizard. Biol J Linn Soc. 2012;105(2): 382–94. doi: 10.1111/j.1095-8312.2011.01798.x

[38] Mastretta-Yanes A, Moreno-Letelier A, Piñero D, Jorgensen TH, Emerson BC. Biodiversity in the Mexican highlands and the interaction of geology, geography and climate within the Trans-Mexican Volcanic Belt. J Biogeogr. 2015;42(9):1586–600. doi: 10.1111/jbi.12546

[39] Bryson RW, Murphy RW, Lathrop A, Lazcano-Villareal D. Evolutionary drivers of phylogeographical diversity in the highlands of Mexico: A case study of the *Crotalus triseriatus* species group of montane rattlesnakes. J Biogeogr. 2011; 38(4):697–710. doi: 10.1111/j.1365-2699.2010.02431.x

[40] Bryson RW, García-Vázquez UO, Riddle BR. Diversification in the Mexican horned lizard *Phrynosoma orbiculare* across a dynamic landscape. Mol Phylogenet Evol. 2012;62(1):87–96. doi: 10.1016/j.ympev.2011.09.007

[41] Nosil P. Ecological Speciation. Oxford University Pres. 2015.

[42] Anacker BL, Strauss SY. The geography and ecology of plant speciation: Range overlap and niche divergence in sister species. Proc R Soc B Biol Sci. 2014;281(1778). doi: 10.1098/rspb.2013.2980

[43] Mallet J. Hybridization as an invasion of the genome. Trends in Ecology and Evolution. 2005;20(5):229–37.doi: 10.1016/j.tree.2005.02.010

[44] Snow R. Continental climate and continentality. In: Encyclopedia of Earth Sciences Series. Dordrecht: Springer. 2005.

[45] Nybom H. Comparison of different nuclear DNA markers for estimating intraspecific genetic diversity in plants. Mol Ecol. 2004;13(5):1143–55. doi: 10.1111/j.1365-294X.2004.02141.x

[46] Miranda F. Anales del Instituto de biología de la Universidad nacional de

México. An del Inst Biol la Univ Nac México. 1955;26:79, f. 1–3.

[47] Schlechtendal DFL. Botanische Zeitung. Berlin.1864;22:144.

[48] Newton AC, Gow J, Robertson A, Williams-Linera G, Ramírez-Marcial N, González-Espinosa M, *et al*. Genetic variation in two rare endemic mexican trees, *Magnolia sharpii* and *Magnolia schiedeana*. Silvae Genet. 2008;57(6): 348–56.

[49] Sexton JP, Hangartner SB, Hoffmann AA. Genetic isolation by environment or distance: Which pattern of gene flow is most common? Evolution (N Y). 2014;68(1):1–15. doi: 10.1111/evo.12258

[50] Chen Y, Chen G, Yang J, Sun W. Reproductive biology of *Magnolia sinica* (Magnoliaecea), a threatened species with extremely small populations in Yunnan, China. Plant Divers. 2016;38 (5):253–8. doi: 10.1016/j.pld.2016.09.003

[51] Oppel S, Mack AL. Bird assemblage and visitation pattern at fruiting *Elmerrillia tsiampaca* (Magnoliaceae) trees in Papua New Guinea. Biotropica. 2010;42(2):229–35. doi: 10.1111/j.1744-7429.2009.00572.x

[52] José AC, Da Silva EAA, Davide AC, Melo AJS, Toorop PE. Effects of drying rate and storage time on *Magnolia ovata* Spreng. seed viability. Seed Sci Technol.2011;39(2):425–34. doi: 10.15258/sst.2011.39.2.14

[53] Gómez Restrepo ML. Fenología reproductiva de especies forestales nativas presentes en la jurisdicción de Corantioquia, un paso hacia su conservación. Medellín: CORANTIOQUIA; 2010. 132 p.

[54] Lamarck J-B de M. Encyclopédie méthodique: botanique, vol 3. chez Panckoucke; 1789. 52–53 p.

[55] Núñez-Cruz A, Meave JA, Bonfil C. Reproductive phenology and seed germination in eight tree species from a seasonally dry tropical forest of Morelos, Mexico: Implications for Community-Oriented Restoration and Conservation. Trop Conserv Sci. 2018; 11. doi: 10.1177/1940082917749946

[56] Bentos T V., Mesquita RCG, Williamson GB. Reproductive Phenology of Central Amazon Pioneer Trees. Trop Conserv Sci. 2008;1(3):186–203. doi: 10.1177/194008290800100303

[57] Cseke LJ, Kaufman PB, Kirakosyan A. The Biology of Essential Oils in the Pollination of Flowers. Nat Prod Commun. 2007;2(12):1317–36. doi: 10.1177/1934578X0700201225

[58] Pergentino de Sousa D. Bioactive Essential Oils and Cancer [Internet]. de Sousa DP, editor. Cham: Springer International Publishing; 2015. doi: 10.1007/978-3-319-19144-7

[59] Robustelli della Cuna FS, Calevo J, Bari E, Giovannini A, Boselli C, Tava A. Characterization and Antioxidant Activity of Essential Oil of Four Sympatric Orchid Species. Molecules. 2019;24(21):3878. doi: 10.3390/molecules24213878

[60] Can Baser K, Buchbauer G. Handbook of Essential Oils: Science, Technology, and Applications [Internet]. Can Baser K, Buchbauer G, editors. CRC Press; 2009. Available from: https://www.taylorfrancis.com/books/9781420063165

[61] Tisserand R, Young R. Essential Oil Safety: A Guide for Health Care Professionals. 2nd ed. Elsevier - Health Sciences Division; 2014. 784 p.

[62] Azuma H, Toyota M, Asakawa Y. Intraspecific Variation of Floral Scent Chemistry in *Magnolia kobus* DC. (Magnoliaceae). J Plant Res. 2001;114 (4):411–22. doi: 10.1007/PL00014006

[63] Shen Y, Li C, Zhou S, Pang E, Story D, Xue C. Chemistry and Bioactivity of Flos Magnoliae, A Chinese Herb for Rhinitis and Sinusitis. Curr Med Chem. 2008;15(16):1616–27. doi: 10.2174/092986708784911515

[64] Park C, Park S-Y, Lee S, Kim J, Park S. Analysis of Metabolites in White Flowers of *Magnolia denudata* Desr. and Violet Flowers of *Magnolia liliiflora* Desr. Molecules. 2018;23(7):1558. doi: 10.3390/molecules23071558

[65] Sarker S, Maruyama Y. Magnolia The genus *Magnolia*. London, UK: Taylor & Francis; 2002. 191 p.

[66] Shalisko V, Vázquez-García JA, Muñiz-Castro MA, Hernández-Vera G, Ortega-Peña AS, Dahua-Machoa A; Osorio-Muñoz EA, Jacobo-Pereira C. Supplementary materials for the chapter How to save endangered magnolias? From population biology to conservation action: the case of allopatric radiation in western Mexico, Mendeley Data, v2. [Internet]. 2021. Available from: http://dx.doi.org/10.17632/jkx452rt9p.2

[67] Linnaeus C. Systema naturae. Editio Dec. Vol. 2. Stockholm Laurentii Salvii; 1759. 1082 p.

[68] Azuma H, Rico-Gray V, García-Franco JG, Toyota M, Asakawa Y, Thien LB. Close Relationship between Mexican and Chinese *Magnolia* (Subtropical Disjunct of Magnoliaceae) Inferred from Molecular and Floral Scent Analyses. Acta Phytotaxon Geobot. 2004;55(3):167–80.

[69] Azuma H, Thien LB, Kawano S. Molecular Phylogeny of *Magnolia* (Magnoliaceae) Inferred from cpDNA Sequences and Evolutionary Divergence of the Floral Scents. J Plant Res. 1999;112(3):291–306. doi: 10.1007/PL00013885

[70] Buckle J. Basic Plant Taxonomy, Basic Essential Oil Chemistry, Extraction, Biosynthesis, and Analysis. In: Clinical Aromatherapy [Internet]. Elsevier; 2015. p. 37–72. Available from: https://linkinghub.elsevier.com/retrieve/pii/B9780702054402000036

[71] Harper JL. Population Biology of Plants. London: Academic Press; 1977. 892 p.

[72] Baskin CC, Baskin JM. Seeds: Ecology, biogeography, and, evolution of dormancy and germination. London: Academic Press. 2014.

[73] Vázquez-Yanes C, Orozco-Segovia A. Physiological Ecology of Seed Dormancy and Longevity. In: Tropical Forest Plant Ecophysiology. Boston: Springer. 1996. p.535–58.

[74] Corral-Aguirre J, Sánchez-Velásquez LR. Seed ecology and germination treatments in *Magnolia dealbata*: An endangered species. Flora Morphol Distrib Funct Ecol Plants. 2006;201(3):227–232. doi: 10.1016/j.flora.2005.07.004

[75] Barbour JR. *Magnolia* L. Bonner FT, Karrfalt RP, editors. Woody Plant Seed Management. Washington: USDA. 2008. p.700–5.

[76] Jacobo-Pereira C, Romo-Campos R, Flores J. Germinación de semillas de *Magnolia pugana* (Magnoliaceae), especie endémica y en peligro de extinción del occidente de México. Bot Sci. 2016;94(3):575–84. doi: 10.17129/botsci.512

[77] Vásquez-Morales SG, Sánchez-Velásquez LR. Seed ecology and pre-germinative treatments in *Magnolia schiedeana* Schlecht, an endangered species from México. J Food, Agric Environ. 2011;9(1):604–8.

[78] Fernando MTR, Jayasuriya KMGG, Walck JL,Wijetunga ASTB. Identifying dormancy class and storage behaviour of champak (*Magnolia champaca*) seeds,

an important tropical timber tree. J Natl Sci Found Sri Lanka. 2013;41(2):141–6. doi: 10.4038/jnsfsr.v41i2.5708

[79] Iralu V, Upadhaya K. Dormancy, storability, and germination of seeds in *Magnolia punduana* (Magnoliaceae). Botany. 2016;94(10). doi: 10.1139/cjb-2016-0056

[80] Vásquez-Morales SG, Ramírez-Marcial N. Seed germination and population structure of two endangered tree species: *Magnolia perezfarrerae* and *Magnolia sharpii*. Bot Sci. 2019;97(1):2–12. doi: 10.17129/botsci.1977

[81] Saldaña-Acosta A, Zuloaga-Aguilar MS, Jardel-Peláez EJ. Germinación de *Acer skutchii* Rehder y *Magnolia iltisiana* Vázquez en la reserva de la Biosfera Sierra de Manantlán, Jalisco, México. For Veracruzana. 2001;3(2):1–8.

[82] Bewley JD, Bradford KJ, Hilhorst HWM, Nonogaki H. Seeds: Physiology of development, germination and dormancy, 3rd edition. New York: Springer. 2013. 392 p.

[83] Ogawa M, Hanada A, Yamauchi Y, Kuwahara A, Kamiya Y, Yamaguchi S. Gibberellin biosynthesis and response during Arabidopsis seed germination. Plant Cell. 2003;15(7):1591–604. doi: 10.1105/tpc.011650

[84] ISTA. International Rules Seed Test. 2016. 384 p. doi: 10.15258/istarules.2016.F

[85] Zuccarini JG. Plantarum novarum. Königlich Bayerischen Akademie der Wissenschaften; 1836. 2:315–380.

[86] Vázquez-García JA, Gómez-Domínguez H, López-Cruz A, Espinosa-Jiménez JA, Sahagún-Godínez E, Muñiz-Castro MA. *Magnolia perezfarrerae*, a new species and a key to Mexican species of Magnolia (section *Talauma*, subsection *Talauma*, Magnoliaceae). Bot Sci. 2013; 91(4):417–25. doi: 10.17129/botsci.421

[87] Gallegos-Mendoza SM, Ortega-Peña AS, Jacobo-Pereira C, Vazquez-García JA. Pruebas de germinación en semillas de *Magnolia pacifica* (A. Vazquez) y *M. vallartensis* (A.Vázquez & Muñiz-Castro). In Guadalajara, México: Consortium and Symposium for the Conservation of Neotropical Magnolias; 2019. 1 p.

[88] Pierre JBL. Flore forestière de la Cochinchine. 1880;1:pl. 3.

[89] Candiani G, Galetti M, Cardoso VJM. Seed germination and removal of *Michelia champaca* L. (Magnoliaceae) in eucalypt stands: the influence of the aril. Rev Árvore. 2004;28(3):327–32. doi: 10.1590/S0100-67622004000300002

[90] Flores J, Arredondo A, Jurado E. Comparative seed germination in species of *Turbinicarpus*: An endangered cacti genus. Nat Areas J. 2005;25(2).

[91] Dirr MA, Heuser CW. The Reference Manual of Woody Plant Propagation: From Seed to Tissue Culture. Athens: Varsity Press. 1987.

[92] Rogers BM, Jantz P, Goetz SJ. Vulnerability of eastern US tree species to climate change. Glob Chang Biol. 2017;23(8):3302–20. doi: 10.1111/gcb.13585

[93] Garavito NT, Newton AC, Golicher D, Oldfield S. The relative impact of climate change on the extinction risk of tree species in the montane tropical Andes. PLoS One. 2015;10(7):e0131388. doi:10.1371/journal.pone.0131388.

[94] Beck HE, Zimmermann NE, McVicar TR, Vergopolan N, Berg A, Wood EF. Present and future Köppen-Geiger climate classification maps at 1-km resolution. Sci Data. 2018;5(1): 180214. doi:10.1038/sdata.2018.214

[95] Thuiller W, Lavorel S, Araújo MB, Sykes MT, Prentice IC. Climate change threats to plant diversity in Europe.

PNAS. 2005; 102(23):8245–50. doi: 10.1073/pnas.0409902102

[96] Randin CF, Engler R, Normand S, Zappa M, Zimmermann NE, Pearman PB, et al. Climate change and plant distribution: Local models predict high-elevation persistence. Glob Chang Biol. 2009;15(6):1557–69. doi: 10.1111/j.1365-2486.2008.01766.x

[97] Vázquez-García JA, Shalisko V, Barragán-Reynaga DM, Villalobos-Arámbula AR, Anguiano M. Probable habitat disappearance in climate change scenarios for recently discovered Ecuadorian endemic *Magnolia mercedesiarum*. Serna M, editor. Aportes para la gestión del riesgo y cambio climático. 1ra ed. 2018;15–33.

[98] Vázquez-García JA, Neill DA, Shalisko V, Arroyo F, Merino-Santi RE. *Magnolia mercedesiarum* (subsect. *Talauma*, Magnoliaceae): a new Andean species from northern Ecuador, with insights into its potential distribution. Phytotaxa. 2018;348(4):254–68. doi: 10.11646/phytotaxa.348.4.2

[99] Hirzel AH, Le Lay G. Habitat suitability modelling and niche theory. Journal of Applied Ecology. 2008;45(5): 1372–81. doi: 10.1111/j.1365-2664.2008.01524.x

[100] Franklin J. Mapping species distributions: Spatial inference and prediction. Cambridge University Press; 2010. 320 p.

[101] Araújo MB, Peterson AT. Uses and misuses of bioclimatic envelope modeling. Ecology. 2012; 93(7):1527–39. doi: 10.1890/11-1930.1

[102] CONABIO, National Geographic Society, California Academy of Sciences. Naturalista.

[103] Fick SE, Hijmans RJ. WorldClim 2: new 1-km spatial resolution climate surfaces for global land areas. Int J Climatol. 2017;37(12):4302–15. doi: 10.1002/joc.5086

[104] Swart NC, Cole JNS, Kharin VV, Lazare M, Scinocca JF, Gillett NP, et al. The Canadian Earth System Model version 5 (CanESM5.0.3). Geosci Model Dev. 2019;12:4823–73. doi: 10.5194/gmd-12-4823-2019, 2019

[105] Eyring V, Bony S, Meehl GA, Senior CA, Stevens B, Stouffer RJ, et al. Overview of the Coupled Model Intercomparison Project Phase 6 (CMIP6) experimental design and organization. Geosci Model Dev. 2016;9:1937–58. doi: 10.5194/gmd-9-1937-2016, 2016

[106] Talluto M V, Boulangeat I, Vissault S, ThuillerW, Gravel D. Extinction debt and colonization credit delay range shifts of eastern North American trees. Nat Ecol Evol. 2017;1:1–6.

[107] Rivers M, Beech E, Murphy L, Oldfield S. The red list of Magnoliaceae-revised and extended. First edit. Botanic Gardens Conservation International; 2016. 63 p.

[108] Cires E, De Smet Y, Cuesta C, Goetghebeur P, Sharrock S, Gibbs D, et al. Gap analyses to support ex situ conservation of genetic diversity in *Magnolia*, a flagship group. Biodivers Conserv. 2013;22:567–90. doi: 10.1007/s10531-013-0450-3

[109] IUCN Standards and Petitions Committee. Guidelines for using the IUCN Red List categories and criteria. Vers 14. [Internet]. 2019. Available from: http://www.iucnredlist.org/documents/RedListGuidelines.pdf

[110] KEWRoyal Botanic Gardens. GeoCAT Geospatial Conservation Assessment Tool. IUCN, KEWRoyal Botanic Garden, Vibrant, Cordis, European Union & NERC. 2019.

[111] Secretaria de Medio Ambiente y Recursos Naturales. NOM-059-

SEMARNAT-2010. Diario Oficial de la Federación 2010.

[112] Cicuzza D, Newton A, Oldfield S. The Red List of Magnoliaceae. Cambridge, UK: Fauna & Flora International; 2007.

[113] Vázquez García JA, Vargas Rodríguez YL, Aragon Cruz F. Descubrimiento de un bosque de *Acer-Podocarpus-Abies* en el municipio de Talpa de Allende, Jalisco, México. Bol del Inst Bot IBUG. 2000;7:159–83.

[114] Vargas-Rodriguez YL, Platt WJ, Vázquez-García JA, Boquin G. Selecting relict montane cloud forests for conservation priorities: The case of Western Mexico. Natural Areas J. 2010; 30(2):156–73. doi: 10.3375/043.030.0204

[115] Fenner M, Thompson K. The ecology of seeds. Cambridge University Press. 2005. 260 p.

[116] Venable DL, Brown JS. The selective interactions of dispersal, dormancy, and seed size as adaptations for reducing risk in variable environments. Am Nat. 1988;131(3): 360–384.

[117] Calderón Sáenz E. Horticultura de *Magnolias* para la Conservación. In: BGCI, editor. Tepatitlán, México: Neotropical Magnolia Conservation Consortium; 2019. p. 32.

[118] Gardiner JM. Magnolias. Chester, Connecticut: Globe Pequot Press. 1989. 144 p.

[119] Figlar RB. Grafting *Magnolia* on *Liriodendron*. J Am Magnolia Soc. 1984; 20(2):21–4.

[120] Biedermann IEG. Factors affecting establishment and development of *Magnolia* hybrids in vitro. Acta Hortic.1987;212:625–30. doi: 10.17660/ActaHortic.1987.212.104

[121] Tobe JD. In vitro growth of *Magnolia grandiflora* L. cv. Bracken's Brown Beauty. J Magnolia Soc. 1990;26:4–8.

[122] IUCN. Guidelines for Re-introductions. Gland, Switzerland: UICN/SSC; 1998. 10 p.

[123] IUCN/SSC. Guidelines for Reintroductions and Other Conservation Translocations. Ver 1.0. Gland, Switzerland: IUCN Species Survival Commission; 2013. viiii+ 57 pp.

[124] Sánchez-Velásquez LR, Pineda-López M del R, Vásquez-Morales SG, Avendaño-Yáñez M de la L. Ecology and conservation of endangered species: The case of magnolias. In: Quinn M, editor. Endangered Species: Threats, Conservation and Future Research. Nova Sciences Publisher. 2016. p. 64–84

[125] Vázquez-García JA, Muñiz-Castro MÁ, Padilla-Lepe J, Ávila MGP, Hernández GN, González REM. *Populus luzae* (salicaceae), a new species of white poplar endemic to the western transmexican volcanic belt, in Zapopan, Jalisco, Mexico. Phytotaxa. 2017;328(3): 243–56. doi: 10.11646/phytotaxa.328.3.3

[126] Pérez De La Rosa JA, Gernandt DS. *Pinus vallartensis* (Pinaceae), a new species fromWestern Jalisco, Mexico. Phytotaxa. 2017;331(2):233–42. doi: 10.11646/phytotaxa.331.2.7

[127] Zabalgoitia A, Figueroa DS, Muñiz-Castro M. A new species of *Miconia* (Melastomataceae) endemic to western Jalisco, Mexico. Phytotaxa. 2020;432: 01–10. doi: 10.11646/phytotaxa.432.1.1

[128] SEMARNAT. Documentos relacionados al Programa para otorgar subsidios de conservación y aprovechamiento sustentable de la vida silvestre nativa. 2020.

[129] CONANP. Pago por Servicios Ambientales en Áreas Naturales Protegidas. 2010.

[130] Lascuráin M, List R, Barraza L, Díaz Pardo E, Gual Sill F, Maunder M, et al. Conservación de especies ex situ. In: CONABIO, editor. Capital natural de México. México: CONABIO; 2009. p. 517–44.

[131] Botanic Gardens Conservation International. Global Survey of Ex-situ Magnoliaceae Collections. Richmond, UK: BGCI; 2008.

[132] Candolle AP. Regni vegetabilis Systema naturale vol 1. Paris: Treuttel et Würtz; 1817.745 p.

[133] Vázquez-García JA, Domínguez-Yescas R, Pedraza-Ruiz R, Sánchez-González A, Muñiz-Castro MÁ. *Magnolia rzedowskiana* (Magnoliaceae), a new species from the section *Macrophylla* in the central part of the Sierra Madre Oriental, Mexico. Acta Bot Mex. 2015;112:19–36.

Chapter 3

Propagation of *Podophyllum hexandrum* Royale to Enhance Production of Podophyllotoxin

Utkarsha Srivastava and Hemant Sood

Abstract

Also known as the Himalayan mayapple, *Podophyllum hexandrum* is a succulent erect herb, glabrous, up to 30 cm tall with creeping long knotty rhizome. The plant produces podophyllotoxin, an anticancer metabolite, and hence can also be used for the treatment of cancer. The roots have also reported anticancer lignans, including podophyllotoxin and berberine. The root is harvested in the autumn and either dried for later use or the resin is extracted. Due to the plant has been overexploited due to medicinal properties and is now enlisted as an endangered species. Therefore, there is a need to grow this plant at a greater scale so as to utilize its medicinal potential to the fullest. Unsuccessful attempts have been made to grow the pant in vitro. Hence, this problem needs to be countered and methods to increase the metabolite production by the plants are also needed in order to maximize the utilization of its medicinal properties. This review focuses on providing solutions to the researchers to develop new techniques to grow the plant in vitro as well as ex situ and also gives an insight on the various methods that have been proved fruitful for increasing the production of podophyllotoxin in *P. hexandrum*.

Keywords: *Podophyllum hexandrum*, anticancer, podophyllotoxin, in vitro propagation, ex situ propagation

1. Introduction

The Himalayan range in India is recognized as a mega hotspot for biodiversity [1]. **Table 1** gives a list of the different types of plants found in this region, which helps to understand the importance of the Himalayan region in terms of endemism of plant species [2]. Of these floral collections, 1748 species have been identified to be of medicinal importance both traditionally and also as modern phytotherpeutics [3]. These 1748 species have been utilized for various purposes (**Table 2**) [3, 4], 121 of which have been categorized to be rare-endangered species [5]. Besides using these local floras for their own survival in form of food, medicine, agricultural tools, and fodder, the local communities also trade these endemic plant species and their traditional knowledge for generating income [3]. According to a survey conducted for the prioritization of medicinal plants by the All India Trade Survey, the demand of these plants for their medicinal value has increased by 50% while their availability has noted a decrease of 25% [6].

S. No.	Flora	No. of species	Percentage of endemism (%)
01	Angiosperms	8000	40
02	Gymnosperms	44	16
03	Pteridophyte	600	25
04	Bryophytes	1737	33
05	Lichen	1159	11
06	Fungi	6900	27

The table also gives the percentage of different floras that are endemic to that area, thereby suggesting the importance of the Indian Himalayan region in terms of biodiversity.

Table 1.
List of different types of flora inhabiting the Indian Himalayan region.

S. No.	No. of species	Use
01	1748	Medicinal plants
02	675	Edible wild plants
03	118	Essential oils yielding medicinal plants
04	279	Species used as fodder
05	155	Species worshiped as sacred plants
06	121	Rare and endangered species

Table 2.
The number of different species found in the Himalayan region has diverse uses which cater to the needs of both the traditional local communities and the industrially significant medicinal market.

The term *Podophyllum* is derived from an ancient Greek word meaning foot leaf. It is also known as mayapple since the fruit ripens in the months of spring. The genus *Podophyllum* is globally represented by three species, the *P. hexandrum*, *P. peltatum*, and *P. sikkimensis*. The *P. hexandrum*, which grows in the Himalayan regions of Asian subcontinent, is commonly known as the Indian mayapple. *P. peltatum* is found distributed in regions of Atlantic North America and is popularly called as the American mayapple [7]. *P. hexandrum* is found to grow normally in the sub-Himalayan regions of India, Pakistan, Bhutan, Afghanistan, China, and Taiwan. In India, it largely covers the states from Jammu and Kashmir to Uttarakhand and some parts of Northeastern India due to its specific environmental requirements. **Table 3** gives a state-wise distribution of this plant species in India [8].

Podophyllum is a green herbaceous plant belonging to the family of Berberidacea of the order Ranunculales. Indian *Podophyllum* is a native of the Himalayan region and grows favorably in the temperate and subalpine regions having well-drained, humus-rich soil conditions. In Ayurvedic terminology, it is referred to as Vanyakarkati. The plant has a perennial rhizome and a succulent stem. The stem has a pair of leaves drooping down from the petiole like umbrellas. The stem grows to a height of approximately 30 cm before the complete development of the leaves. The leaves are generally spotted with a width of nearly 25 cm. The flower is pinkish white in color and appears in May. The fruit ripening occurs in August or September and the fruit is bright orange in appearance with a size comparable to that of a lemon.

Several studies have been conducted to identify the best possible methods of propagation of *P. hexandrum*. According to one such study, the propagation of the

S. No.	State-wise distribution
01	Ladakh—Zanskar and Suru valley
02	Jammu and Kashmir—Kashmir region
03	Himachal Pradesh—Lahaul, Spiti, Kangra, Chamba, and Kinnaur
04	Uttarakhand—Kumaun and Garhwal region (2000–4000 m above MSL)
05	Sikkim
06	Arunachal Pradesh

Table 3.
State-wise distribution of P. hexandrum *in India.*

seeds of the plant in an off-site environment under controlled conditions allowed the seedlings to grow a year faster as compared to that when grown in field. The search is on to identify and establish such effective ex situ methods of propagation as these along with several other modes of propagation as in vitro and in vivo methods can help conserve the genetic diversity of the plant besides providing a substantial number of transplants to go back into the wild to combat the vulnerability of overharvesting without compromising on its industrial demand [9]. There is very less literature available showing the growth of *Podophyllum hexandrum* using micro-propagation techniques.

Use of bioreactors for the growth and production of podophyllotoxin has not yet been evaluated properly owing to the lack of genetic data about the plant's genes involved in metabolite production. The technical know-how about the factors affecting biosynthetic pathway and podophyllotoxin production has not been explored and no significant literature is available. This review focuses on the optimization of culture conditions for the propagation of *Podophyllum hexandrum* and obtains conditions for the enhanced production of podophyllotoxin (**Figure 1**).

Kingdom: Plantae

Subkingdom: Tracheiobionta

Superdivision: Spermatophyta

Division: Magnoliophyta

Class: Magnoliopsida

Subclass: Magnolide

Order: Ranunculales

Family: Berberidaceae

Genus: *Podophyllum* L.

Species: *Podophyllum hexandrum* Royale

Figure 1.
Taxonomic classification of P. hexandrum.

2. Podophyllotoxin—resin from *Podophyllum hexandrum*

Podophyllum hexandrum has been extensively studied primarily for its medicinal properties that are contained in its resin extract, podophyllotoxin. It occurs extensively in the roots and the rhizome of the plant species. The content of podophyllotoxin is also dependent on the growth conditions of the plant, including the environmental factors as soil pH, rainfall, temperature, humidity, etc. [10]. This resin can be extracted from both the species of the plant, *P. hexandrum* (Indian) and *P. peltatum* (American), although it has been well established that the yield of the Indian plant is greater than that of its American counterpart. A number of different components were later isolated from the podophyllin that was isolated both from the American and Indian species [11].

α- peltatin β- peltatin	Isolated from *P.peltatum*
4'- demethylpodophyllotoxin Picropodophyllin-β-D-glucoside	Isolated from *P.hexandrum*

2.1 Phytochemical profile of podophyllotoxin

Podophyllotoxin is a member of the aryltetralin lignans family according to its chemical structure. It is a product of phenylpropane units which are coupled together by β-carbons in their side chain. **Table 4** gives a list of the compounds characterized from both the *Podophyllum* resin as well as *Podophyllum* species [12]. **Figure 2** depicts the chemical structure of the podophyllotoxin in both 2D and 3D formats [13].

Chemical formula of podophyllotoxin: $C_{22}H_{22}O_8$.

Molecular weight of podophyllotoxin: 414.4 g/mol.

A cycle of seven precursors is involved in the production of this resin naturally [14]. Podophyllotoxin is the most active naturally occurring cytotoxic product, hence it is used as a principle ingredient in the preparation of its semisynthetic derivatives that function as cytostatics and are therefore used in the treatment of several types of cancer. The major anticancer drugs obtained from this toxin are etoposide and teniposide. **Figures 3** and **4** give the chemical structure of etoposide [15] and teniposide [16]. Podophyllotoxin inhibits the assembly of the microtubule, thereby inhibiting the process of cell division. It is also reported to have certain antiviral activities by interfering with certain vital viral processes [17].

The quantity of the resin collected is variable with the season and site of collection. The maximum yield of toxin can be obtained in May, when the plant is to flower and decreases in near November 7%, when the plant is in fruiting stage. Also, the yield obtained is higher from the young rhizomes. As the rhizomes mature, the amount of podophyllotoxin accumulation decreases. Assessing the difficulties in the stating and execution of an appropriate methodology for obtaining higher yields of this toxin, besides increasing the numbers of this species in the wild, the present review aims to study and analyze some of the various methods that have been performed to achieve these objectives.

2.2 Medicinal value of podophyllotoxin

The rhizome of the plant contains a resin, known generally and commercially as Indian *Podophyllum* Resin, which can be processed to extract podophyllotoxin or podophyllin, a neurotoxin. Podophyllotoxin is the major lignan present in the resin

Compound	
Lignans	**Flavinoids**
1. Podophyllotoxin	1. Quercetin
2. α-peltatin	2. Kaempferol
3. β-peltatin	3. Isorhamnetin
4. 4′-demethylpodophyllotoxin	4. Quercetin 3-galactoside
5. Desoxypodophyllotoxin	
6. Dehydropodophyllotoxin	
7. Sikkimotoxin	
8. Podophyllotoxin glucoside	
9. Picropodophyllinglucoside	
10. α-peltatinglucoside	
11. β-peltatinglucoside	
12. 4′-demethylpodophyllotoxinglucoside	

Table 4.
List of different compounds that can be obtained from podophyllotoxin or the Podophyllum species. These compounds can either be lignans or flavinoids chemically.

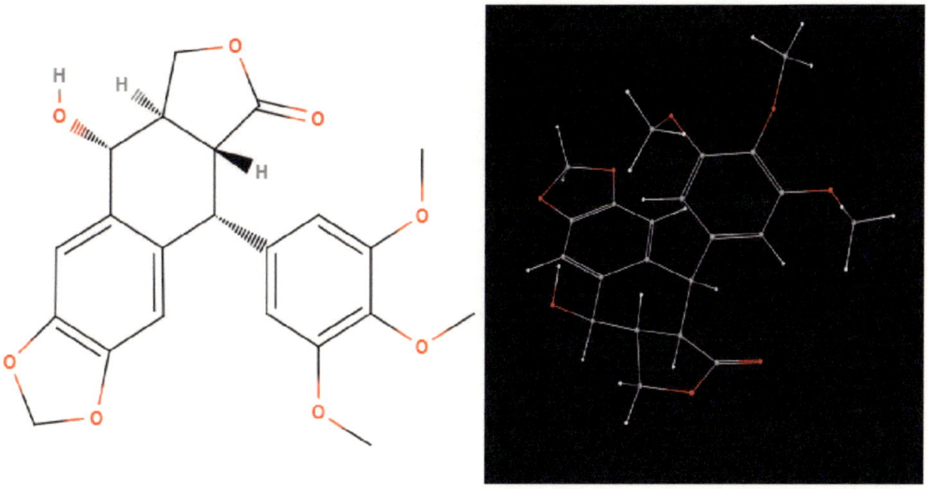

a b

Figure 2.
Chemical structure (a) of the resin podophyllotoxin and its 3D representation in form of wireframe (b).

and is a dimerized product of the intermediates of the phenylpropanoid pathway. The starting material of etoposide (vepeside), an FDA approved anticancer drug, is podophyllotoxin and has been used to treat testicular cancer as well as lung cancer by inhibiting replication of cancer cells. Podophyllotoxin finds use as a precursor for the semisynthetic topoisomerase inhibitors in the treatment of leukemias, lung and testicular cancers, and dermatological disorders like warts, rheumatoid arthritis, and psoriasis. It also has numerous applications in modern medicine by virtue of its free radical scavenging capacity. An extract of *P. hexandrum* has been shown to provide approximately 80% whole-body radioprotection in mice [17]. Twenty-five percent solution of *Podophyllum* resin is efficacious and a cost-effective treatment with minimal side effects for HIV-related oral hairy leukoplakia, which is a symptom-free lesion

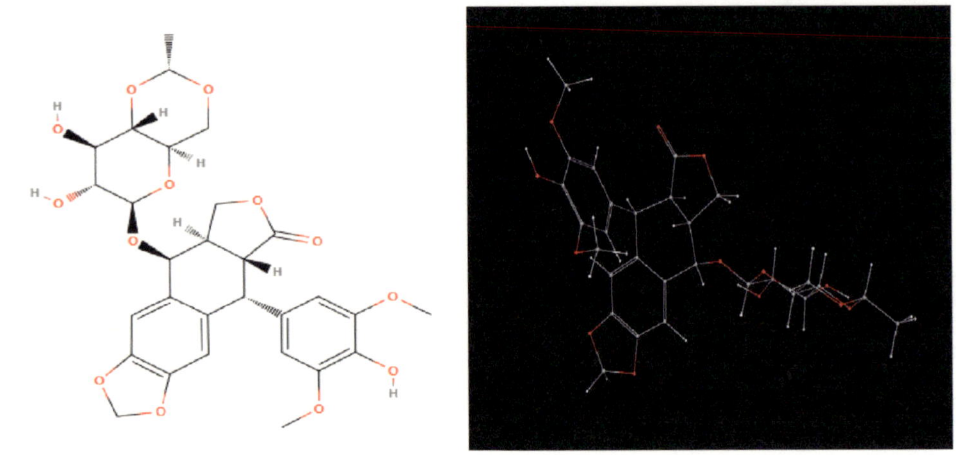

a b

Figure 3.
The chemical structure of anticancer drug etoposide in 2D (a) and 3D (b) forms.

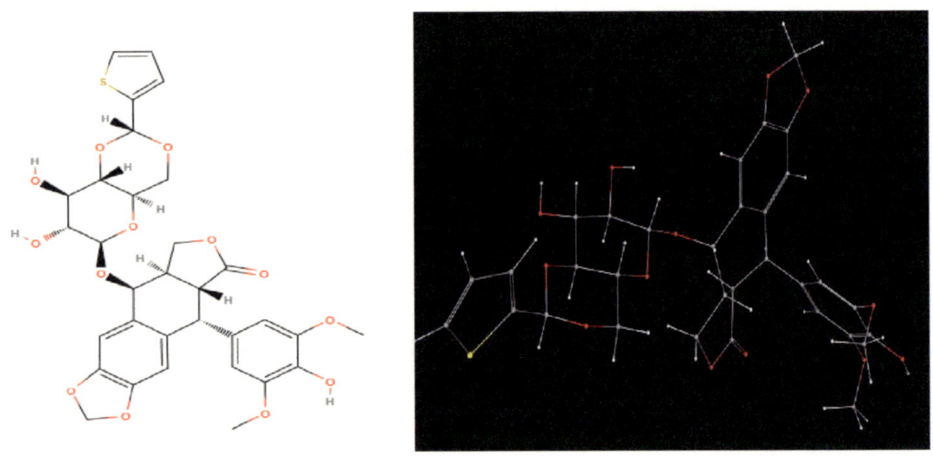

a b

Figure 4.
Chemical structure of anticancer drug teniposide in 2D (a) and 3D (b) forms.

[18]. Another in vitro study showed podophyllotoxin as a promising cytotoxin against a set of human cancer cell lines: HL-60, A-549, HeLa, and HCT-8. PTOX was also found to activate proapoptotic endoplasmic reticulum stress signaling pathway [19].

Etoposide, teniposide, and etopophos are the different anticancer drugs derived from podophyllotoxin. These compounds are topoisomerase II inhibitors. Topoisomerase II enzyme is essentially required to cleave the double-stranded DNA and to seal it again after unwinding. It is crucial in the process of DNA replication and repair. Etoposide and other derivatives stabilize the DNA-topoisomerase II complex in a way so that resealing of DNA strands becomes impossible. Cells that are duplicating their DNA in the S phase and preparing for mitosis are very sensitive for this mechanism. The overall effect of these anticancer drugs is the arrest of the cells in late S or early G2 phase of the cell cycle [20–22].

Apart from being an important anticancer compound, podophyllotoxin is also found to possess various other important medicinal properties, some of which include:

Protection against radioactivity: Several researches have confirmed that various extracts of podophyllotoxin including chloroform, methanolic and hydro-alcoholic extracts provided 70–95% protection against radioactivity [23–25].

Antifungal activity against *Aspergillus niger* and *Candida albicans* [26].

The dichloromethane extract of this compound is investigated to possess insecticidal activity [27].

Traditional application: Used as an antihelminthic by Native Americans. In India, the aqueous extracts of the roots have been used as cathartic and also to cure ophthalmia [28].

2.3 Production of podophyllotoxin

Podophyllotoxin is chemically a member of the lignin group of compounds. Lignans are dimerziation products of two phenylpropane units linked by the β-carbon atom of the side chain [29]. Most of the pathways proposed involve phenolic oxidative coupling of C6-C3 monomers via schikimic acid pathway. Production of optically active lignan dimmers is an enzyme-controlled reaction [30]. A series of compounds of considerable commercial and medicinal interest as clinically useful anticancer drugs are formed by the reductive dimerziation of cinnamic acid or cinnamic alcohols [17]. **Figure 5** gives the biosynthetic pathway for the production of podophyllotoxin in *Podophyllum* species [31].

The synthesis of the derivative compounds as secondary metabolites occurs due to the diversified properties of the ring structures. The pathway for the biosynthesis of podophyllotoxin starts from coniferyl alcohol which is converted into pinoresinol in the presence of an oxidant through a series of reactions that involve the dimerization of a stereospecific reaction intermediate. Complete and conclusive knowledge on this pathway is still not available and research is on to incur more information about the genes and transcription factors that may be involved in the regulation of this pathway.

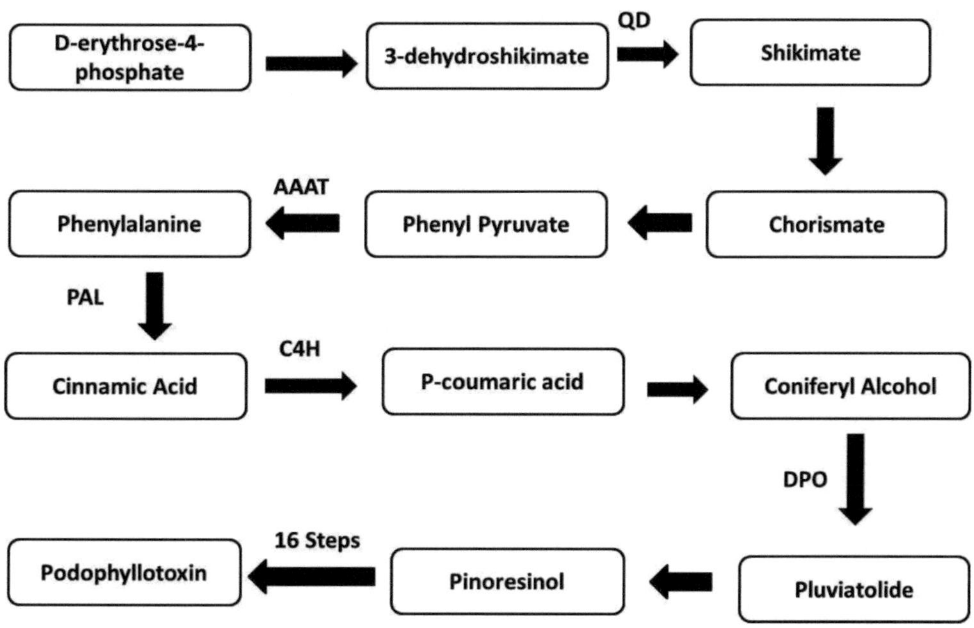

Figure 5.
Biosynthetic pathway for the production of podophyllotoxin (QD: Quinate dehydrogenase; AAAT: Aromatic amino acid transaminase; PAL: Phenylalanine ammonia lyase; C4H: Cinnamate-4-hydroxylase; and DPO: Dirigent protein oxidase) [30, 32, 33].

2.4 Limitations in the propagation of *Podophyllum hexandrum*

Since *Podophyllum hexandrum* is an endangered species and its resin has wide medical applications, there is a necessity to propagate the plant. But the process of propagation of the plant under both natural and laboratory conditions has some strict limitations which restrict the process. One of the major problems for cultivation of this plant is its long juvenile phase and poor fruit setting ability. Also, its seeds take a long period to germinate [34]. The plant has a low capacity of regeneration in natural environment and with the overexploitation of the plant coupled with the accelerated rate of destruction of its natural habitat, it is becoming extremely difficult to revive the plant in the wild.

Since the plant has strict requirement for conditions regarding the growth of the plant in the fields, therefore it is not amenable for cultivation as an agricultural crop, especially in the lowland areas which constitute a major percentage of land in India. Although in vitro approaches for the propagation of this plant with enhanced production of podophyllotoxin have long been studied and tested in various researches, the lack of complete knowledge of the pathway involved in the biosynthesis of podophyllotoxin has made these approaches limited to a small group of growth culture media and supplements which might enhance its production as well as its propagation. Cell and tissue culture techniques, though have shown some hope, are commercially not feasible, and therefore, cannot be used. Several studies are now being conducted on alternative approaches to optimize the culture conditions for the growth of this plant along with enhancement in the yield of podophyllotoxin to find a suitable technique that is both commercially feasible and experimentally reproducible.

3. In vitro method of propagation

Techniques of plant tissue culture have long been explored as instruments for the mass production of many overexploited and medicinally important plants as well as secondary metabolites. In vitro plant, cell, and organ cultures have been considered more feasible and amenable as compared to whole plants for the production of secondary metabolites since the plants are cultivated in simple and well-defined media under controlled conditions and they are independent of the natural environment for their growth and survival.

3.1 Somatic embryogenesis

The study isolated embryogenic callus from zygotic embryos and placed in 30-ml MS media supplemented with NAA and PVP. They kept the culture in complete darkness in rotary shaker (100 rpm) at 25 ± 2°C. After establishing the optimum strength for the MS basal media, embryogenic calli were cultured on MS media (0.75 strength) supplemented with 3 g/l PVP and varying concentrations of sucrose, glucose, fructose, and mannose. Thirty proliferated somatic embryos were cultured on 0.75 strength of MS basal media supplemented with 3 g/l of PVP and varying concentrations of ABA. The cultures were incubated at 25 ± 2°C for 16-h photoperiod and analyzed after 2 weeks. Matured somatic embryos were transferred to 0.75 strength of MS basal media supplemented with PVP and varying concentrations of GA3. Somatic embryos germinated on GA3 were dried and ground to fine powder and podophyllotoxin was extracted. Quantification of podophyllotoxin using water system with PDA detector at a wavelength of 250 nm was performed. Relative amounts of podophyllotoxin were calculated by comparing the peaks from the chromatogram [22]. 2, 4-D, and NAA were seen to have profound effect on the callus growth. 1 mg/l of 2, 4-D in combination with 3 g/l of PVP gave the best results for culture establishment. A change in osmotic pressure directly affects the development of embryos. Best results of callus growth were obtained for 0.75 strength of MS basal media. It also had higher podophyllotoxin content. Sucrose was found to be the best carbon source, and 4% sucrose with 0.75 strength of MS basal media gave better results. 1 mg/l ABA concentration showed efficient maturation and plants showed better podophyllotoxin content at this concentration. The study concluded that best suspension cultures may be obtained for 0.75 strength of MS basal media supplemented with 1 mg/l 2, 4-D, and GA3, each with 4% sucrose and incubation at 25 ± 2°C.

3.2 Precursor feeding

It has been reported that upon using coniferin as a precursor for the podophyllotoxin, production increased by 12.8 times. The problem with this technique is that coniferin is not commercially available. Therefore, Lin et al. devised a technique of coculturing of *Linum flavum* hairy roots and *Podophyllum hexandrum* cell suspensions in 2003. In this study, *Linum flavum* hairy roots and *P. hexandrum* cell suspensions were used to build a coculture system for the in vitro production of podophyllotoxin. *Agrobacterium rhizogenes* strains, LBA9402 and TR105, were used to initiate hairy roots from seedlings of *L. flavum*. The roots were maintained with liquid MS media supplemented with sucrose of conc. 30 g/l at a pH of 5.9. The roots were incubated in flasks containing 25-ml MS media in a rotary shaker running at 100 rpm in the dark. [35]. *L. flavum* hairy roots were cocultured with *P. hexandrum* cell suspension cultures in a dual 500-ml shaker flask with the bottom side openings linked by a 4–5 cm length of 3 mm silicone tubing. The culture was incubated at 25°C in dark in a rotary shaker running at 100 rpm. *Linum flavum* hairy roots and *P. hexandrum* cell suspensions were cultured separately in two 2-L bioreactors. 5 g FW of 3-week old *L. flavum* root was inoculated directly into 1.8 L of LS medium and the airflow rate was set at 80–100 cm3/min for maintaining DO tension above 85%. 360 ml of 3-week old suspension cultures of *P. hexandrum* were inoculated in LS medium and total volume was made up to 1.8 L and the airflow rate was set at 120–150 cm3/min to maintain the dissolved oxygen tension above 80% air saturation. Medium exchange between the two plants was started 12 days after the inoculation. Cultures were harvested after 29 days and medium samples were analyzed periodically. The results from the study concluded that the dual bioreactor containing the coculturing of the two plants showed a better podophyllotoxin concentration per biomass (mg-1 dry weight) of 0.062 as compared to 0.032 mg-1 dry weight in single reactor. The concentration of coniferin was also found to increase in the reactor containing both the plants.

3.3 Production through hairy root cultures

The study showed that strains of *A. rhizogenes* used for embryo transformation in *P. hexandrum* produced transformed calli. HPLC profiling of these transformed calli revealed that the culture contained three times more podophyllotoxin in contrast to controls [36]. In this study, the seeds of *P. hexandrum* were rinsed in Tween 20 and surface sterilized with 0.2% mercuric chloride for 10 min and soaked in water for a day. Dissected embryos were cultured on MS medium and incubated in continuous light at 25 ± 2°C. *Agrobacterium rhizogenes* strains viz. 15,834, Aq, and K599 were grown on nutrient agar at 29°C and cultured in YMB liquid medium for 48 h. Different explants were examined for the induction of hairy root cultures.

Ten- to 15-day old aseptically growing embryos of *P. hexandrum* (in two sets) were wounded and incubated in acetosyringone (25 mM) in combination with lO mM glucose, 5 mM morpholino ethane sulphonic acid (MES), and 150 mM NaCl and were incubated for 20 min in 48-h old cultures of the respective *A. rhizogenes* strains. The embryos were then transferred to MS basal media containing acetosyringone (50 pM) with and without 2, 4-D, BAP and were incubated in two sets, one in light and the other in dark. The growth of transformed and control cultures was monitored after culturing 40 mg of inoculum in 50 ml of medium in 250-ml flasks in triplicates. The cell suspension culture was harvested every 3 days up to the 12th day and every 2 days after 12 days and the increase in weight was recorded. For podophyllotoxin production, tissue (both transformed and control) was air dried, weighed, and powdered. Extraction was done by treating the callus with methanol for 8 h in a soxhlet apparatus at 60°C. The methanolic extract was concentrated in a rotavapor. The residue was dissolved in methanol (AR) prior to analysis. Podophyllotoxin content was analyzed by HPLC. It was concluded from the study that the strains of *A. rhizogenes* used, namely A4 and 15,834, showed fast-growing calli at the site of infection. Hairy root phenotype was not observed despite the addition of acetosyringone alone or in combination with glucose, MES, and NaCl. Approximately, threefold increase in the podophyllotoxin content was observed as compared to control cultures. HPLC analysis indicated a maximum of 0.7% podophyllotoxin in cell suspension cultures derived from callus lines transformed with *Agrobacterium rhizogenes* strains, A4 and 15,834, while the control calli gave a maximum of 0.2% podophyllotoxin only. Highest accumulation of podophyllotoxin in cell suspension cultures was obtained during the stationary phase up to 18 days after which it declined.

3.4 Biotransformation

Biotransformation approach of podophyllotoxin production has helped to develop the derivatives of this resin, which have enhanced the anticancer properties coupled with the antimitotic activity of podophyllotoxin. So, Rajesh et al. initiated the *Agrobacterium*-mediated biotransformation of *P. hexandrum* for increased production of podophyllotoxin [35]. Mature seeds of the plant were collected from its natural habitat, washed with running tap water, and then rinsed with 0.1% (v/v) Teepol solution. The seeds were then sterilized with 70% (v/v) ethanol for 1 min followed by 0.1% mercuric chloride for 10 min and were finally rinsed several times with sterile double-distilled water. The seeds were stored in flasks containing 30 ml of sterile double-distilled water for a day on an orbital shaker running at 120 rpm. Three strains of *A. tumefaciens*, LBA 4404, EHA 101, and bEHA 105 containing the pCAMBIA 2301 binar vector having nptII and gusA genes were used. Both these genes are controlled by CaMV 35S promoter and poly (A) terminator. The cocultivated embryogeniccalli were washed and later inoculated in MS basal media supplemented with 150 mg/l kanamycin and 200 mg/l timentin and then were incubated for 6 weeks at 25 ± 2°C under a 16-h photoperiod. The surviving embryogenic calli were separated and subcultured onto fresh selection media. The matured somatic embryos were germinated for 2 weeks on a germination medium (GM) containing 150 mg/l kanamycin at 25 ± 2°C under a 16-h photoperiod. Rooting plantlets were transferred into paper cups containing perlite, peat moss, and vermiculite (1:1:1 v/v/v) and covered with polythene bags to maintain 80% relative humidity. The results from the study showed that timentin was found to exhibit a better efficiency than cefotaxime at all the concentrations tested. Timentin at 200 mg/l inhibited the growth of the three strains of *Agrobacterium* that were tested. Cefotaxime controlled *Agrobacterium* growth at 300 mg/l at which the rate of somatic embryogenesis was 27.33% (13.66 out of 50 calli responded). Three days of cocultivation proved to be optimal as under these conditions, 65.33% of embryogenic calli (32.66 out of 50) GUS histochemical analysis revealed that the transgene was successfully integrated and expressed in the *P. hexandrum* genome.

4. Ex situ method of propagation

Attempts to obtain podophyllotoxin through cell cultures or chemical synthesis techniques are still far from being economically feasible. The objective of the following study was to enhance the root formation and podophyllotoxin production of *P. hexandrum* cultivated in a glasshouse [37]. Two batches of plants grown for different time periods were obtained from two different regions and stored at 7–8°C in the dark to prevent the formation of shoot. They were later cultivated in glasshouse in the peat-perlite soil (2:1 w/w). For every condition and time point, 15 plants were randomly harvested. The root biomass and podophyllotoxin content of the plants of each temperature group were analyzed at the beginning. The plants were harvested for a period of 20 or 40 days, but all the plants were cultivated for a minimum period of 20 days before giving them the methyl jasmonate treatment. Fifteen plants were immediately harvested for baseline control, 30 plants were sprayed with water (control), and 30 plants were sprayed with 5 l of 1.5 mM methyl jasmonate. After 9 days, 15 plants from each group were harvested for analysis. The plants in the treatment group were sprayed again for 3 consecutive days with 5 l of 3 mM methyl jasmonate each day and harvested for analysis the next day. Roots from each plant were collected, rinsed with tap water, and dried for 18 h at 40°C. They were later pooled in groups of three, ground and stored at room temperature in closed containers in the dark. 10 ml of methanol was added to 1 g of the plant material. The sample was vortexed at 2500 rpm and incubated at 65°C in a water bath for 10 min. Then the mixture was centrifuged at 4°C at 2400 g for 10 min. The supernatant was separated and transferred to a fresh tube. This extraction process was performed five times. The podophyllotoxin content was determined by HPLC analysis and the samples were stored at 4°C before analysis. Podophyllotoxin is stable in the refrigerator at 4°C for at least 3 months and at 25°C. The results of the study showed a higher concentration of podophyllotoxin in MeJ-treated plants (30 mg/g) as compared to MeJ-deficient plants (18 mg/g).

5. Latest advancements in production of podophyllotoxin and growth of *Podophyllum hexandrum*

5.1 Role of endophytes in the production of podophyllotoxin

Endophytes are the group of microorganisms that are found to colonize the interior of the plant irrespective of the type of association they have to maintain with the host. Although this group includes a number of species of microbes, it is less explored. They have great potential of application in agriculture, bioremediation, medicine, etc. [38]. A study reported the isolation of an endophytic fungus, *Fusarium solani*, from the roots of *P. hexandrum*, which was found to synthesize podophyllotoxin at a rate of 29.0 μg/g on a dry weight basis [39]. The results were confirmed by HPLC and mass spectroscopy techniques. Upon isolation of the fungal hyphae from the roots of *P. hexandrum*, selection of the fungal species was done by assessing the presence of podophyllotoxin in the three strains that had been isolated, after which the rDNA analysis confirmed the fungus under study to be *Fusarium solani*. Shake flask experiments were performed, the results of which indicated that maximum biomass production was obtained on the 10th day of growth while the maximum yield in podophyllotoxin was found around the 8th day of growth.

5.2 Identification of key transcription factors involved in regulation of biosynthetic pathway for podophyllotoxin production in *P. hexandrum*

Although the exact pathway of biosynthesis of podophyllotoxin is still not known, information about the regulatory components of this pathway is also unavailable. Considering the importance of transcription factors and their role in upgrading the industrial synthesis of podophyllotoxin, the present study worked to identify different transcription factors that might be involved in regulating the pathway of podophyllotoxin biosynthesis [40]. The study worked to identify various classes of transcription factors via the mining of transcriptomes of *Podophyllum* species and validation of these factors by qRT-PCR analysis coupled with the analysis of podophyllotoxin content from the different tissues of *P. hexandrum*. Extensive survey of available literature revealed that four transcription families (TFs), basic leucine zipper (bZIP), myeloblastosis (MYB), WRKY, and basic helix-loop-helix (bHLH) were involved in the regulation of phenylpropanoid pathway in several species of plants. It was therefore hypothesized that these TFs might also be involved in the regulation of biosynthesis of podophyllotoxin in the species of *Podophyllum*. Two distinctive transcripts were identified which encoded for bZIP and MYB TFs in the rhizomes of *P. hexandrum* and which were associated with podophyllotoxin content. Upon quantifying the content of podophyllotoxin and analyzing the comparative expression between the high (2.51%) versus the low (0.59) content of podophyllotoxin accessions, the results showed a 0.04- to ~16-fold increase in the transcripts of the transcription factors, further supporting the involvement of the identified TFs with the content of podophyllotoxin. For *P. hexandrum*, the highest transcript abundance was observed for bZIP (19.60-fold) in the rhizome showing 2.51% of podophyllotoxin as compared to the shoots which showed only 0.01% of the resin. In silico analysis of putative promoter regions of the genes associated with this pathway in other species of plants have shown the presence of certain sequence elements for MYB and WRKY TFs, which suggested their involvement in regulating the production of podophyllotoxin. Abundance of the transcript was evaluated with respect to the transcription families using fragments per kilobase of transcripts per million mapped reads (FPKM) and qRT-PCR-based transcript by in silico techniques. The values of TFs FPKM (fragments per kilobase of transcripts per million mapped reads) ranged between 0.0014–12.01 and 0.014–1162.01 in the transcriptomes correlating with shoots and rhizomes of *P. hexandrum* and *P. peltatum*, respectively. Gene expression pattern was observed through two different platforms and the results were in synchrony with each [31].

6. Conclusion

From the study, it can be concluded that growth of *Podophyllum hexandrum* is difficult in vitro, but the plant can be grown successfully by somatic embryogenesis provided that media and other culture conditions are optimized in a stepwise manner. The traditional culturing techniques do not tend to increase the podophyllotoxin production, whereas its production can be increased by coculturing of hairy roots of the plant along with *L. flavum*. This process is called is precursor feeding while molecular techniques such as genetic engineering of the plant with the help of

Agrobacterium tumefaciens have also showed an increase in podophyllotoxin production. The other methods to increase metabolite production include the growth of hairy root culture and spraying of methyl jasmonate in the plants under greenhouse conditions. There are certain gaps which have not been taken care of like there is very less literature available showing the growth of *Podophyllum hexandrum* using micropropogation techniques. The genetic data about the plant's genes involved in metabolite production are not easily available. Use of bioreactors for the growth and production of podophyllotoxin has not yet been evaluated properly. There is further in-depth study required to be done on *Podophyllum hexandrum* before we can begin the large-scale podophyllotoxin production from the plant.

The formation of State Medicinal Plant Board ensures increased cultivation and conservation of the medicinally important endangered plants by providing appropriate funding to conserve these species [41]. Although the efforts of agrotechnological innovation and biotechnological processes cannot be undermined, only a few of these endemic plant species are under cultivation majorly by the efforts of local farmers and NGOs [3].

Acknowledgements

The authors would like to acknowledge the support and constant guidance of Dr. Hemant Sood whose invaluable input to this review chapter has made it worth a read. The authors would also like to thank the Department of Biotechnology and Bioinformatics, Jaypee University of Information Technology, for providing them with the opportunity to work on this project.

The authors also wish to acknowledge the support of Mr. Priyansh Srivastava for the help that he extended during the writing of this chapter.

Conflict of interest

The authors declare that they have no conflict of interest whatsoever.

Author details

Utkarsha Srivastava* and Hemant Sood
Jaypee University of Information Technology, Solan, Himachal Pradesh, India

*Address all correspondence to: utkarsha0587@gmail.com

References

[1] Myers N, Muttermeier RA, Muttermeier CA, da Fonseca ABG, Kent J. Biodiversity hotspots for conservation priorities. Nature. 2000;**403**:853-858

[2] Singh DK, Hajra PK. Floristic diversity. In: Gujral GS, Sharma V, editors. Biodiversity Status in the Himalaya. New Delhi: British Council; 1996. pp. 23-38

[3] Samant SS, Dhar U, Palni LMS. Medicinal Plants of Indian Himalaya: Diversity Distribution Potential Values. Nainital: Gyanodaya Prakashan; 1998

[4] Samant SS, Dhar U. Diversity, endemism and economic potential of wild edible plants of Indian Himalaya. International Journal of Sustainable Development and World Ecology. 1997;**4**:179-191

[5] Nayar MP, Sastry ARK. Red Data Book of Indian Plants, Vol. I–III. Calcutta: Botanical Survey of India; 1987, 1988, 1990

[6] Anonymous. Appendix I, II and III to Convention on International Trade in Endangered Species of Wild Fauna and Flora. Arlington VA: US Fish & Wildlife Services; 2001. p. 32

[7] Chatterjee R. Indian Podophyllum. Economic Botany. 1952;**6**(4):342-354

[8] Chaurasia OP, Bhatt BB, Singh DV. In: Jina PS, editor. Ethnobotany of Ladakh Himalaya, In:Ladakh: Past & Present. New Delhi: Gyan Publishing House; 2000

[9] Kharkwal A, Kushwaha R, Prakash O, Ogra R, Bhattacharya A, Nagar P, et al. An efficient method of propagation of *Podophyllum hexandrum*: An endangered medicinal plant of the Western Himalayas under ex situ conditions. Journal of Natural Medicines. 2008;**62**(2):211-216

[10] Alam M, Naik P. Impact of soil nutrients and environmental factors on podophyllotoxin content among 28 *Podophyllum hexandrum* populations of northwestern Himalayan region using linear and nonlinear approaches. Communications in Soil Science and Plant Analysis. 2009;**40**(15-16):2485-2504

[11] Hartwell JL. IX-peltatin, a new compound isolated from *Podophyllum peltatum*. Journal of the American Chemical Society. **69**:29r8 (r947)

[12] Hartwell JL, Schrecker AW. The chemistry of *Podophyllum*. Fortschr Chem Org Naturst. 1958; **15**:83-166. DOI: 10.1007/978-3-7091-7162-2_3. PMID: 13597975

[13] National Center for Biotechnology Information. PubChem Compound Summary for CID 10607, Podofilox. PubChem. Available from: https://pubchem.ncbi.nlm.nih.gov/compound/Podofilox [Accessed: 19 August 2020]

[14] van Uden W, Pras N, Malingré T. On the improvement of the podophyllotoxin production by phenylpropanoid precursor feeding to cell cultures of *Podophyllum hexandrum* Royle. Plant Cell, Tissue and Organ Culture. 1990;**23**(3):217-224

[15] National Center for Biotechnology Information. PubChem Compound Summary for CID 36462, Etoposide. PubChem Available from: https://pubchem.ncbi.nlm.nih.gov/compound/Etoposide [Accessed: 19 August 2020]

[16] National Center for Biotechnology Information. PubChem Compound Summary for CID 452548, Teniposide. PubChem. Available from: https://

pubchem.ncbi.nlm.nih.gov/compound/Vumon [Accessed: 19 August 2020]

[17] Giri A, Giri C, Dhingra V, Narasu M. Enhanced Podophyllotoxin production from agrobacterium rhizogenes transformed cultures of *Podophyllum hexandrum*. Natural Product Letters. 2001;**15**(4):229-235

[18] Giri A, Lakshmi Narasu M. Production of podophyllotoxin from *Podophyllum hexandrum*: A potential natural product for clinically useful anticancer drugs. Cytotechnology. 2000;**34**(1/2):17-26

[19] Ardalani H, Avan A, Ghayour-Mobarhan M. Podophyllotoxin: A novel potential natural anticancer agent. Avicenna journal of phytomedicine. 2017;**7**(4):285-294

[20] You Y. Podophyllotoxin derivatives: Current synthetic approaches for new anticancer agents. Current Pharmaceutical Design. 2005;**11**(13):1695-1717

[21] Witterland A, Koks C, Beijnen J. Etoposide phosphate, the water soluble prodrug of etoposide. Pharmacy World & Science. 1996;**18**(5):163-170

[22] van Uden W, Pras N, Visser J, Malingré T. Detection and identification of Podophyllotoxin produced by cell cultures derived from Podophyllum *hexandrum* Royle. Plant Cell Reports. 1989;**8**(3):165-168

[23] Goel HC, Arora R, Prasad J. A process for preparation of a radioprotective herbal extract - I and II. Indian Patent Filed, Patent Office. New Delhi, India; 2000

[24] Goel HC, Arora R, Prasad J. A Process for Preparation of a Radioprotective Herbal Formulation. Indian patent filed, Patent Office, New Delhi, India; 2001

[25] Goel HC, Prakash H, Ali A, Bala M. Podophyllum hexandrum modulates gamma radiation-induced immunosuppression in Balb/c mice: Implications in radioprotection. Molecular and Cellular Biochemistry. 2007;**295**:93-103. DOI: 10.1007/s11010-006-9277-5

[26] Wani SA, Shah KW, Ahmad MA. Antifungal activities of methanolic extracts of *Podophyllum hexandrum* and *Rheum emodi* against human pathogenic fungal strains. International Journal of Pharmaceutical Sciences Review and Research. 2013;**19**:56-59

[27] Miyazawa M, Fukuyama M, Yoshio K, Kato T, Ishikawa Y. Biologically active components against *Drosophila melanogaster* from *Podophyllum hexandrum*. Journal of Agriculture and Food Chemistry. 1999;**47**:5108-5110

[28] Qazi PH, Rashid A, Shawl SA. *Podophyllum hexandrum*—A versatile medicinal plant. International Journal of Pharmacy and Pharmaceutical Sciences. 2011;**3**:261-268

[29] Rajesh M, Sivanandhan G, Subramanyam K, Kapildev G, Jaganath B, Kasthurirengan S, et al. Establishment of somatic embryogenesis and podophyllotoxin production in liquid shake cultures of *Podophyllum hexandrum* Royle. Industrial Crops and Products. 2014;**60**:66-74

[30] Jackson D, Dewick P. Aryltetralin lignans from *Podophyllum hexandrum* and *Podophyllum peltatum*. Phytochemistry. 1984;**23**(5):1147-1152

[31] Pawan K et al. Expression analysis of biosynthetic pathway genes vis-à-vis podophyllotoxin content in *Podophyllum hexandrum* Royle. Protoplasma. 2015;**252**(5):1253-1262

[32] Marques J, Kim K, Lee C, Costa M, May G, Crow J, et al. Next Generation Sequencing in Predicting Gene Function

in Podophyllotoxin Biosynthesis. Journal of Biological Chemistry. 2012;**288**(1):466-479

[33] Xia ZQ, Costa MA, Proctor J, Davin LB, Lewis NG. Dirigent-mediated podophyllotoxin biosynthesis in *Linum flavum* and *Podophyllum peltatum*. Phytochemistry. 2000;**55**:537-549

[34] Chu C, Huang C, Chu N. Sensory neuropathy due to Bajiaolian (Podophyllotoxin) intoxication. European Neurology. 2000;**44**(2): 121-123

[35] Nandagopal K, Halder M, Dash B, Nayak S, Jha S. Biotechnological approaches for production of anti-cancerous compounds resveratrol, Podophyllotoxin and Zerumbone. Current Medicinal Chemistry. 2018;**25**(36):4693-4717

[36] Rajesh M, Jeyaraj M, Sivanandhan G, Subramanyam K, Mariashibu T, Mayavan S, et al. Agrobacterium-mediatedtransformation of the medicinal plant *Podophyllum hexandrum* Royle (syn. *P. emodi* wall. Ex Hook.f. & Thomas). Plant Cell, Tissue and Organ Culture (PCTOC). 2013;**114**(1):71-82

[37] Seegers C, Setroikromo R, Tepper P, Horvatovich P, Peters R, Quax W. Methyl jasmonate treatment increases podophyllotoxin production in *Podophyllum hexandrum* roots under glasshouse conditions. Plant and Soil. 2017;**417**(1-2):117-126

[38] Biswas D, Biswas P, Nandy S, Mukherjee A, Pandey D, Dey A. Endophytes producing podophyllotoxin from *Podophyllum* sp. and other plants: A review on isolation, extraction and bottlenecks. South African Journal of Botany. 2020

[39] Nadeem M. Fusarium solani, P1, a new endophytic podophyllotoxin-producing fungus from roots of Podophyllum hexandrum. African Journal of Microbiology Research. 2012;**6**(10)

[40] Kumar P, Jaiswal V, Pal T, Singh J, Chauhan R. Comparative whole-transcriptome analysis in *Podophyllum* species identifies key transcription factors contributing to biosynthesis of podophyllotoxin in *P. hexandrum*. Protoplasma. 2016;**254**(1):217-228

[41] Kala CP, Sajwan SB. Revitalizing Indian systems of herbal medicine by the National Medicinal Plant Board through institutional networking and capacity building. Current Science. 2007;**93**(6):797-806

Chapter 4

Noteworthy Threatened Plant Species in the Sahel Region, Nigeria

Abdallah Muhammad Salihu and Rusea Go

Abstract

The loss of biodiversity in Nigeria is intensifying distressingly, which happened to be more prominent in some states across northwestern as well as north eastern geopolitical regions of Nigeria and was threatened by the confrontational effect of desertification. The Sahel region and some part of the savannah region bordered which faced the menace of drought and desertification which lead to so many factors like soil erosion, distraction of the ecosystem entirely. Moreover, the plants destroyed were mostly aromatic plants which served as medicines in curing many ailments and diseases contributed to the economic status of the communities found in the Sahel. The plants species became threatened to the human practices by noticing their effects for their well-being. Consequently, among the species that were threatened include: *Neocarea macrophylla* Prance ex F.White (Gawasa), *Sclerocarya birrea* (A. Rich) Hochst. (Danya), *Detarium microcarpum* Harms. (Taura), *Prosopis africana*, (Kirya), *Acacia africana*, *Acacia nilotica* and *Azadirachta indica*. Hence, the major concern of this chapter is to assess some of the threatened species found in the Sahel region, Nigeria and to highlight the valuable medicinal plants at risk of extinction according to the guidelines of the International Union for the Conservation of Nature (IUCN).

Keywords: African, dessert plants, medicinal plants, threatened, ethnobotany

1. Introduction

The loss of biodiversity in Nigeria is intensifying distressingly ("Assessment of the threats to biodiversity" Convention on Biological Diversity [CBD]), which happened to be more prominent in some states across northwestern as well as north eastern geopolitical regions of Nigeria and was threatened by the confrontational effect of desertification, posing a substantial threat to the well-being of the populace [1]. The Sahel region and some part of the savannah region bordered with Niger Republic faced the menace of drought and desertification that leads to soil erosion and distraction of the ecosystem entirely (**Figure 1**). The confrontational effect is caused by the rampant cutting of trees, loss of ground cover and continuous overgrazing by livestock of the Fulani herdsmen. Moreover, the plants destroyed were mostly aromatic and medicinal plants gathered, traded and used by communities in Sahel as source of livelihood. Consequently, among the species that were threatened include *Neocarea macrophylla* (Sabine) Prance ex F.White

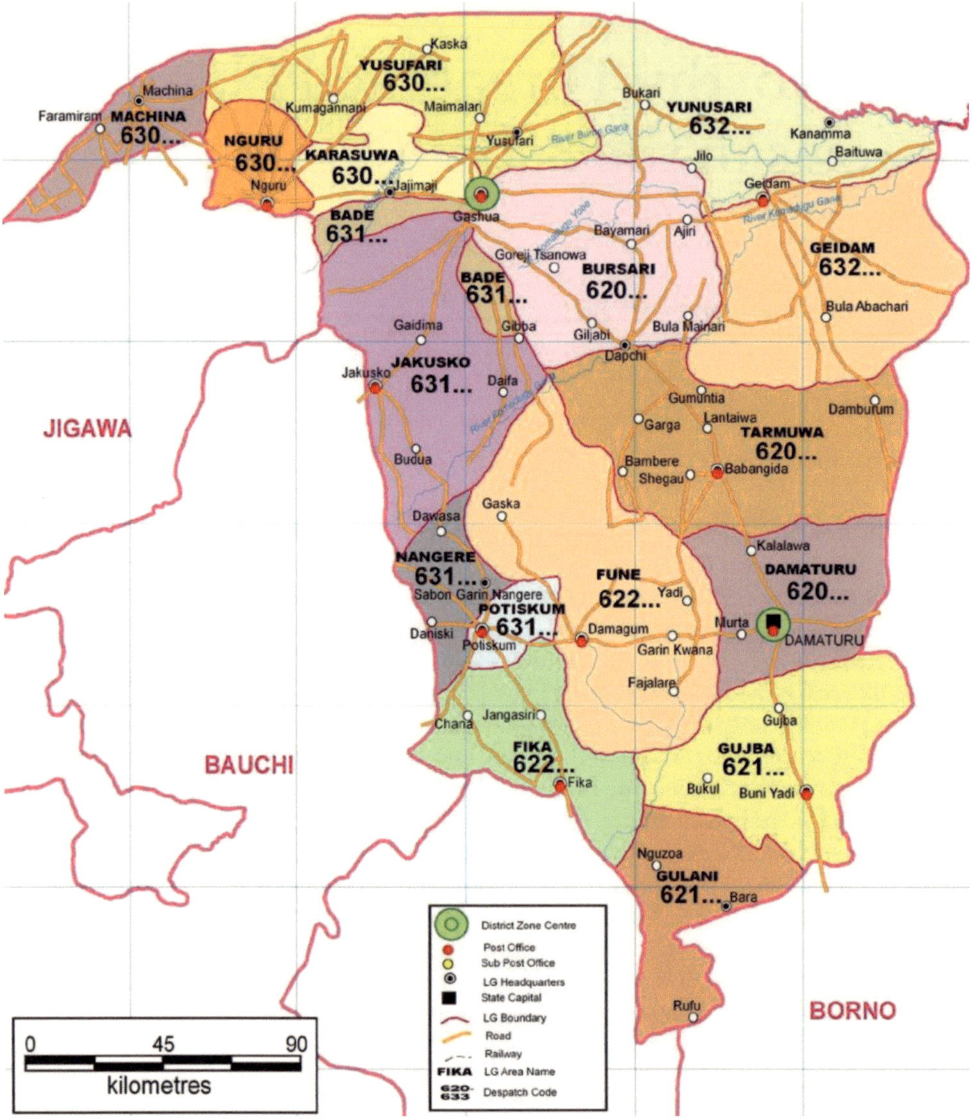

Figure 1.
Map of Yobe State, Nigeria (Sahel region).

(Gawasa), *Sclerocarya birrea* (A. Rich) Hochst. (Danya), *Detarium microcarpum* Harms. (Taura), *Prosopis africana* (Guill. and Perr.) Taub. (Kirya), *Acacia africana*, *Acacia nilotica*, *Azadirachta indica* [2]. Though numerous international and national summit have been conducted towards ways to tackle the menace of desertification but apparently failed in the Sahel regions, notably because majority of the populace relied on the natural resources for their well-being. The genus *Acacia* with about 1380 species is widely distributed from Australia to rest in tropical and subtropical regions of the globe is called "*bagaruwa*" by the Hausa tribe (Sudan and Nigeria). It is used to treat many ailments by utilizing its leaves, stem and pods simply because the abused of many commercially made antibiotics has led to many antibiotic resistances in human pathogens [2]. Consequently, *S. birrea* considered to be native African tree and widely distributed across the continent. Its stem-bark, roots, leaves and fruits are contained a myriad of chemical components, which attributed to its utilization as food and traditional medicine for many ailments. Its fruits were regarded as wild fruits for their uniqueness of nutritional attributes of ascorbic acid and other hydrocarbons. The seeds possessed essential acids and oil used as anti-inflammatory, antidiabetic, analgesic, anti-parasitic, antimicrobial and antihypertensive [3]. Hence, *S. birrea* (marula) is commercially valuable as many tribes in Africa used its stem-bark for making inks. The wood part of "marula" mainly used for dishes, mortars, drums, toys, curios and cosmetics oil [4]. It has become a challenge to Nigerian Ministry of Health over many years because both rural and urban communities opted to medicinal plants as complementary and alternative medium of curing their ailments, and over cutting unsustainably and lack of enforcement caused some species into verge of extinction [5].

1.1 Loss of biodiversity

Evidences showed that the biodiversity has been unprecedentedly lost at a higher rate in many parts of Nigeria, which are mostly human related factors such as industrialization, technological advancements and settlements for urbanization. Biodiversity loss in Nigeria is mainly due to poor law enforcement, high demand for forest products and cultural practices. In addition, clearing of vegetation haphazardly leads to loss in biodiversity. However, biodiversity loss may also due to some agricultural and livelihood practices such as firewood cutting and gathering, bush burning, logging and overgrazing. Since 1990s, some large farm has caused enormous deforestation of the natural habitats depicting as the major source of loss. Thus, drastic depletion of fire-wood species like neem tree, *Acacia nilotica*, *Acacia seyal* was very prominent [6]. Consequently, the Sahel region is characterized by small, widely disseminated, small-leaved trees, mainly *Acacia* species such as *Acacia tortilis subsp. raddiana*, *Acacia seyal*, *Acacia nilotica* and *Acacia senegal*. Equally high percentage of the trees was deciduous in the dry season. Some species were *Anogeissus leiocarpus*, *Sclerocarya birrea* (**Figure 2a**), *Combretum micranthum* and *Guiera senegalensis*. Interpretation of radar imagery distinguishes the following (predominantly) shrub vegetation types: "Dense shrub grassland" and "shrub grassland" which offer a woody cover of shrub like *Combretum* spp., *Guiera senegalensis*, *Acacia* species, *Calotropis senegalensis* and *Boscia senegalensis*, adequately classified them as wooded formations [6].

1.2 Plants species utilization patterns

Northern Sahel region of Nigeria nutritional habits are equivalent with slight variations influenced by traditional backgrounds. All the trees selectively protected in the parklands provide one or combinations of the following concrete and immaterial services: food items (edible fruits, nuts, leaves, flowers, livestock fodder, fuel wood, medicine, esthetics, shade, agricultural tools and cooking utensils, avenues, and other services such as ropes, fibers, tannin, manure, latex gums and oils). Patterns of exploitation are quite similar for the entire region. However, additional trees were protected based on their medicinal values and food, Fire wood (fuel energy and heating) stand equally of importance, with the exception of *Adansonia digitata* and *Calotropis procera*, all species were utilized as fire wood obtained from prunings, complete felling or pollarding [6]. All the constituent parts of the tree featured regular and often in the food security and nutritional resources for the people and with some quantity sole in the market as sources of income. Many of the cattle farmers still rely substantially on those trees for chopped forage, while the fruits, shells were also fed to the livestock. Shade provision was unique with *Parkia biglobosa*, *Adansonia digitata* and *Tamarindus indica* alongside with *Mangifera indica* due to their broad canopies. Currently, income from the sales

Figure 2.
Threatened useful plants from the Sahel region in Nigeria. (a) Sclerocarya birrea, *(b)* Acacia seyal, *(c)* Azadirachta indica, *(d)* Acacia nilotica, *(e)* Balanites aegyptiaca, *(f)* Detarium microcarpum, *and (g)* Tamarindus indica.

of fruits of *Parkia biglobosa*, *Vitellaria paradoxa, Adansonia digitata, Tamarindus indica*, *Balanites aegyptiaca* and *Acacia seyal* (**Figure 2b**) accounted for 30-45% of the farmers' income. Common known exotics like *Mangifera indica*, *Anacardium* *occidentale* and *Azadirachta indica* (**Figure 2c**) were on a balance with most the indigenous trees and were equally threatened. Soil fertility nourishment happened to be secondary role from some of these trees through litter falls. The Sahel savannah covers an area of 31,468 km^2, where some of the species were also threatened in parks and fetish channels arising from remarkable reasons, which bans the cutting down of some species such as *Adansonia digitata*, *Tamarindus indica*, *Acacia senegal* and *Faidherbia albida* [6].

Nevertheless, small number of both species was assessed due to the fact that, many were threatened, deteriorated as a result of human and edaphic factors. Many were not assessed in the IUCN Red list due to the aforementioned factors that pressurized on them to be disappeared or not even listed in the whole list [7]. All stakeholders showed consistent preference for the following species: *Adansonia digitata*, *Acacia seyal*, *Acacia nilotica* (**Figure 2d**), *Acacia senegal*, *Anogeissus leiocarpus*, *Azadirachta indica*, *Anacardium occidentale*, *Balanites aegyptiaca*, *Borassus aethiopum*, *Ceiba pentandra*, *Combretum microcarpum*, *Eucalyptus camaldulensis*, *Diospyros mespiliformis*, *Faidherbia albida*, *Guiera senegalensis*, *Gmelina arborea*, *Hyphaene thebaica*, *Lawsonia inermis*, *Lannea acida, Khaya senegalensis*, *Moringa oleifera*, *Mangifera indica*, *Parkia biglobosa*, *Piliostigma thonningii, Pterocarpus erinaceus*, *Pinus* spp., *Salvadora perica*, *Senna siamea*, *Tamarindus indica*, *Vitellaria paradoxa*, *Vitex doniana* [8]. Highly valued naturalized exotic fruits trees including *Mangifera indica*, *Psidium guajava*, *Anacardium occidentale*, *Azadirachta indica* and *Citrus* spp. Nevertheless, *Acacia senegal* has been exploited commercially for gum Arabic with plantations existing in Yobe state, other plantations were mainly for afforestation and shelterbelts planted up with exotics including *Pinus* spp., *Eucalyptus* spp., *Gmelina arborea*, *Azadirachta indica* and *Senna siamea* [9, 10].

1.3 Other threatened species

A considerable numbers of indigenous species were reported to have declined in some years back. The focus groups credited the decline of these species due to some activities (land degradation) occurred [11]. Certain percentages have shown that the species faced a lot of problems (Ranging 36%-60%) of all the species found in the Sahel. The common plants species involved across the towns and villages were *Acacia tortilis, Ficus glumosa*, *Balanite aegyptiaca* and *Sclerocarya birrea,* whereas, some disappeared including *Albizia chevalieri*, *Acacia senegal, Ficus thonningii* and *Khaya senegalensis,* many factors have contributed towards the disappearance of those species. Furthermore, some exotic species were present and also served many roles in the environment without the much existence of those threatened species especially *A. indica* which has been proclaimed to cause a setback on the farm produce due to some negative impacts on the farm lands. The level of poverty has tremendously attributed towards the negative actions across the threatened plants species, for the search of food, shelter and other agricultural practices and very poor planting and management of trees across the region [12].

2. Methodology

Survey was conducted across the region in collating the reliable information in conjunction with search engines for the peer reviewed journals and books in getting sound and reliable information with regards to the threatened species found in Sahel region. Information on the uses of plants especially those with medicinal properties were carried out by using a survey form and interviewing traditional medicine practitioners, herbs gatherers and sellers. Images of the plants mentioned were taken in the natural habitat and voucher specimens were collected and preserved following the standard herbarium technique.

2.1 Collection of plant materials and identification

Plant species collected were organized as complete herbarium specimens and identified as outline by the rules of herbaria. Plant materials were identified

No.	Botanical name	Family	Common name	Local name	Habit	Part(s) used	Voucher no.
1.	*Adansonia digitata*[***]	Malvaceae	Baobab tree	Kuka	Tree	Stem, leaves	MSA36
2.	*Anogieeossus leocarpa*[**]	Combretaceae	African birch	Marke	Tree	Stem, leaves	MSA29
3.	*Azadirachta indica*[*]	Meliaceae	Neem	Darbejiya	Tree	Stem, leaves	MSA348
4.	*Balanites aegyptiaca*[*]	Zygophyllaceae	Desert tree	Aduwa	Tree	Stem	MSA359
5.	*Cassia singueana*[**]	Fabaceae	White or winter cassia	Runhu	Herb	Leaves	MSA316
6.	*Detarium microcarpum*[*]	Fabaceae	Sweet dear	Taura	Tree	Leaves	MSA71
7.	*Gueira senegalensis*[****]	Combretaceae	Senegal Gueira	Sabara	Shrub	Leaves, root, gall	MSA32
8.	*Khaya senegalensis*[**]	Meliaceae	Mahogany	Madaci	Tree	Stem	MSA116
9.	*Leptadania hastate*[***]	Apocynaceae	Kayila	Yadiya	Herb	Leaves	MSA248
10.	*Piliostigma reticulata*[***]	Fabaceae	Camel's foot	Kargo	Tree	Stem	MSA72
11.	*Prosopis africana*[***]	Fabaceae	Iron tree	Kirya	Tree	Stem	MSA193
12.	*Psidium guajava*[**]	Myrtaceae	Guava	Goba	Tree	Leaves	MSA336
13.	*Sclerocarya birrea*[****]	Anacardiaceae	Marula	Danya	Tree	Stem	MSA435
14.	*Senna italica*[***]	Fabaceae	Italian thorn	Filasko	Herb	Leaves	MSA68
15.	*Tamarindus indica*[***]	Fabaceae	Tamarind	Tsamiya	Tree	Stem, leaves	
16.	*Vachellia nilotica*[****]	Fabaceae	Arabic gum	Bagaruwa	Tree	Pods	MSA74
17.	*Zizyphus mauritania*[***]	Ramnaceae	Chinese Apple	Magarya	Tree	Leaves	MSA186

[****]*Most cited plants.*
[***]*Second most cited plants.*
[**]*Partially cited plants.*
[*]*Least cited plants.*

Table 1.
Medicinal plant species collected from the survey area with their medicinal values.

and authenticated by a plant taxonomist in person of Dr. Yusuf Nuhu, from the Department of Plant Biology, Bayero University Kano, Nigeria. All identified plants specimens were given vouchers number as outline in (**Table 1**).

2.2 Data organization

The collected data was quantified using some quantitative indices such as Informant consensus factor (ICF), Relative frequency of citation (RFC) and fidelity level pinpointed by [13, 14].

2.3 Relative frequency of citations

It was used to authenticate the local benefit of each species in the study areas. Its index was determined by dividing the number of informants cited useful species (FC) by the total number of informants participated in the survey (N), as viz.: RFC = FC/N [15].

3. Results

The collated plants were identified and authenticated in line with the responses from the respondents. They were given voucher numbers as outlined in **Table 1** below, as well as the citation frequencies were procured based on the procured data as shown in **Table 2**.

No.	Species	Family	FC	RFC
1.	*Adansonia digitata*	Malvaceae	6	0.06
2.	*Anogieeossus leocarpa*	Combretaceae	4	0.04
3.	*Azadirachta indica*	Malvaceae	1	0.01
4.	*Balanites aegyptiaca* (**Figure 2e**)	Zygophyllaceae	1	0.01
5.	*Cassia singueana*	Fabaceae	4	0.04
6.	*Detarium microcarpum* (**Figure 2f**)	Fabaceae	1	0.01
7.	*Gueira senegalensis*	Combretaceae	33	0.34
8.	*Khaya senegalensis*	Meliaceae	3	0.03
9.	*Leptadania hastate*	Apocynaceae	5	0.05
10.	*Piliostigma reticulata*	Fabaceae	9	0.09
11.	*Prosopis africana*	Fabaceae	9	0.09
12.	*Psidium guajava*	Myrtaceae	2	0.02
13.	*Sclerocarya birrea*	Anacardiaceae	20	0.21
14.	*Senna italica*	Fabaceae	8	0.08
15.	*Tamarindus indica* (**Figure 2g**)	Fabaceae	5	0.05
16.	*Vachellia nilotica*	Fabaceae	26	0.27
17.	*Zizyphus mauritania*	Ramnaceae	6	0.06

Citation frequencies as far the species collated.

Table 2.
Relative frequency citations values of the surveyed medicinal plants in Yobe state, Nigeria.

4. Discussion

Ethnobotanical information gathered were quantitatively analyzed using various quantitative indices which made up of: Relative frequency of citation (RFC) and Frequency of citation (FC) [14, 16]. It has been proved that, the RFC was happened to be (0.25) for the gastrointestinal disorders among the collated medicinal plants [13], which conforms to the present study as pinpointed in **Table 2**. It has been revealed that, 80% of African populace relied on herbal/traditional medicine to treat many ailments due to the in availability and affordability of the commercial orthodox. In Northern Nigeria, people make used of combined plants formulations to get rid of gastric ulcer problem, which include *A. nilotica, Balanites aegyptiaca* and *Ziziphus abyssinica* extracts, have been concluded and validated that, using combined plants formulations showed sound activity than using individual plants in treating ulcer [17]. *Gueira senegalensis* has been in records and used by traditional practitioners as medicine in many parts of West Africa including Nigeria. The latter confirms the present study as shown in **Table 1**. Its leaves were taken for many purposes like; pulmonary and respiratory problems, colic and diarrhea, syphilis, beriberi, leprosy, rheumatism, diuresis, impotence and expurgation [16]. Furthermore, its roots and bark have been used in treating diarrhea, dysentery and abdominal pain which was found worthy that, the whole plant has pharmacological attributes by serving as antimicrobial, analgesic, anti-inflammatory and also applicable to address animals' ailments and their supplements as diet [18]. Over utilization made some to be in an extinction due to the lack of enforced law [5], which is similar to the current study as shown in (**Figure 3**) the effect of over destruction or deforestation of plant species.

Figure 3.
Major caused of depleting plant resources in the Sahel region. (a) Demand for timber and (b) bush fire.

5. Case study: neem tree

5.1 An overview

Neem (*Azadirachta indica*) is a common tree throughout Sahel, for a worthy reason. It grows dynamically under semi-arid and humid tropical conditions as well as responds to numerous needs. Its dispersed canopy provides shade from the tropical sun. In the Sahel region, neem tree of many uses including as source of fire wood, it protects millet, sorghum and other crops from desert wind and insect pests as it possessed insecticidal properties, and myriads of medicinal importance. Neems in the Sahel region of both Nigeria and Niger do lacked some minerals in certain period in a year; this resulted to yellowish leaves and stunted growth in the end, appeared unhealthy. Many trees have died due to the lack of those vital minerals [6]. Moreover, Neem trees are attractive broad-leaved evergreens that can grow up to 30 m tall and 2.5 m in girth. Their dispersal divisions' form rounded crown as much as 20 m across. They continue in leaf except during life-threatening drought, when the leaves may fall off. The short, usually straight trunk has a moderately thick, strongly furrowed bark. The roots enter the soil deeply, at least where the site permits, and, particularly when injured, they produce suckers. This suckering has a habit of to be especially prolific in dry localities. Neem can take significant abuse. For example, it easily withstands pollarding and its topped trunk resprout robustly. The small, white, bisexual flowers are borne in axillary clusters. They attract many bees for honey. Neem honey happened to be popular, and in fact contains no trace of azadirachtin. The fruit is up to 2 cm long, when developed, it became yellow or greenish yellow and comprised a sweet pulp enfolding a seed. It is the kernel that was used most in pest control (The leaves also contain pesticidal ingredients, but as a rule they are much less effective than those of the seed). A neem tree normally start bearing fruit after 3-5 years, became fully productive in 10 years' time and from then, can yield up to 50 kg of fruits annually. It may live for more than two centuries [6].

5.2 Neems characteristics

Shade is not the only appreciated characteristic of neem. It has so many uses and potentialities for future usage, that was why it considered as miraculous, it is used for furniture, fuel and in construction, it also attract bees and honey flavor. Neem has been declared as local pharmacy. In India, people use neem twig to prevent from teeth damage. It also cured skin disorders, enhanced tonic, treats infectious diseases and fevers [19].

5.3 Neem as part of the West African trees

Neem was brought to West African region through Ghana in the year 1919 and 1927. It became familiar and well spread across towns, villages and cities including Sahel regions. Neems have been declined due to some biotic and abiotic factors which lead to its deterioration in the whole region; it served as a great set back in the development of the Sahel region [20]. Moreover, certain factors also lead to neem declination such as; defoliation by insects, drought or an exposure to pollutants at times certain microbes' infestation may lead to decline in the neem population as well, similar case has been reported on *Acacia nilotica* in Sudan which shared same peculiarities with the Sahel parts of Nigeria [19]. Nevertheless, Sahel normally has a very long dry periods, (July–September rainy season), neem has been reported widely adapted to the arid and semi-arid tropics, long-term stress could result from the developing of neem in plantations where trees compete for available moisture, a narrow genetic base with little variability in susceptibility to stress, poor planting techniques and localized soil nutrients deficiencies in combination with a long dry period as well. Furthermore, other factors such as soil compaction also leads to the declination of neem especially areas of much human and livestock activities resulted to some fungi such as *Nigrospora* and *Curvilaria* as well as cruising by some livestock [21].

5.4 Uses of neem tree in the Sahel

Native to India and Myanmar, the neem being a member of the family Meliaceae together with the mahoganies. It does possessed compound leaves of nine to 15 leaflets which are dark green in color. The fruits are yellow-green to green, smooth, olive-shaped and about 2 cm in length, with a sweet pulp enclosing a seed. Consequently, Neems can grow up to 30 m in height and 70 cm in diameter, with broad, dispersal pinnacles that retain their foliage all year round. Neem due to its attributes, it is highly valued [19].

5.4.1 Timber

Neem is a member of the mahogany family. It is moderately heavy, with a specific gravity varying from 0.56 to 0.85 (average, 0.68). With a strong smell when freshly cut. Although simply sawn, worked, refined, and glued, it must be dried wisely as it often splits and warps. It also splits easily and nailed. However, it is widely used in carts, tool handles, and agricultural implements. In South India became very common furniture wood. It is aromatic, attractively spotted, narrowly linked, and then medium to be coarse in texture. Although it lends itself to carving, it does not take a high polish. The timber appeared durable even in exposed situations. It is rarely attacked by termites, its resistant to woodworms, and makes useful fence posts and poles for house construction. Also used as pole wood especially in developing countries; the tree's capability to resprout after cutting and to regrow its canopy after pollarding makes neem highly suited to pole production. In view of the above, those rigorous activities have threatened its population as well [6].

5.4.2 Fuel

Neem became threatened due to its useful oils produced; it usually burnt in lamps throughout many countries. Its wood has long been in the practice of burning for food as well. Furthermore, husk produced by the seeds, mainly employed as fuel. Because of the tree's good growth and valued firewood, it has become the most vital plantation species in northern Nigeria. It is also grown for fuel around large towns. Charcoal made from this neem wood stands excellent quality, with a rich value only somewhat below that of coal from Nigeria's some of the eastern parts. Neem is very common, especially in towns and villages, in the northern regions despite been threatened [6].

5.4.3 Medicinal values of neem

Although masses in India insisted on the effectiveness of neem actions in treating many ailments, the pharmacological properties have hardly been threatened to severe trials with controls. It has been proclaimed that, neem trees have been in practice in curing many diseases most especially the oil extracted from it, but many findings contradict with the claim to be wise enough in curing children's at the tender age [22]. Neem being a Meliaceae family, a famous plant with medicinal attributes since time immemorial. *Azadirachta indica* have a numerous medicinal attributes. Notably, leaf and bark were used in the treatment of gingivitis, periodontitis, sores, boils, enlarged spleen and malarial fever during childbirth, measles, smallpox, head scald, as well as cutaneous affections [23]. Moreover, seed oil from neem and essential oils of both leaves and bark possessed the potentialities of curing many ailments notably; *Mycobacterium* and *Plasmodium* [24]. Neem oil was used as contraceptive (Vaginal infections) and mosquito repellent due to the fact that, nimbin, nimbinin were found and nimbidin inclusive [23].

5.5 Neem needed a tougher law of protection

Neem deterioration is increasingly appeared in large areas of Nigerian states most especially in neighboring countries like; Niger, Cameroun, Chad and Mali as a result of inaction to the problem occurred in the areas by the government and non-governmental agencies that can curve the menace. Certain measures can lead to the solution to the neem declination which will definitely enhance the socioeconomic effects of neem deteriorations [25]. Neem tree served as an insecticide by possessing some pesticidal components, it attacks so many as widely practiced in West Africa. But in some parts of the world, such as India, Far East Asia, it defoliates and also kills the tree. Reports have been documented that an oriental yellow scale damaged several neem trees in across West African countries including North eastern Nigeria and Eastern Niger, which lead to an eminence drought in the Sahel, which turned many neem trees weak and sickly [26].

6. Conclusion

The biodiversity has been lost at a higher rate in most parts of Nigeria especially in Sahel region, where people solely depend on plants to carry out their life activities. Many factors have contributed a lot most importantly human related, such as: medicinal purposes, industrialization, technological advancements and settlements for urbanization. Moreover, direct causes of biodiversity loss in Nigeria made up of; poor law enforcement and weak laws, much demand from forest products, cultural practices which contributed tremendously in cutting down of vegetation and lead to loss in biodiversity as depicted, many plants would likely to be in extinction if care is not much taken. Proper awareness on the sustainable use of these mostly utilized species within the Sahel region should be forcefully and continuously communicated to the communities living of the resource so that they could one day be the custodian and guardian of their forests and natural resources. We believed that effective conservation and sustainable used of natural resources got to be community based, coupled with national and international law to safeguard their livelihood.

Acknowledgements

We really appreciates the Yobe State University and Universiti Putra Malaysia for funding and facilities rendered for the preparation of this publication. We are thankful to the traditional medicine practitioners, herbs gatherers and sellers for their cooperation in sharing their knowledge of plants.

Conflict of interest

The present book chapter contribution has no conflict of interest declared.

Author details

Abdallah Muhammad Salihu[1,2] and Rusea Go[2]*

1 Desert Research Monitoring and Control Centre, Yobe State University, Damaturu, Yobe State, Nigeria

2 Department of Biology, Faculty of Science, Universiti Putra Malaysia, UPM Serdang, Selangor, Malaysia

*Address all correspondence to: rusea@upm.edu.my

References

[1] Abubakar Bello1 Sirajo Jamaladdeen1 Muhammad T. Elder1 Samaila S. Yaradua1 Sulaiman S. Kankara, Nasiru, H. Wagini1 Charles H. Stirton2 Muthama Muasya. Threatened medicinal and economic plants of the Sudan Savanna in Katsina State, northwestern Nigeria. *Bothalia - African Biodiversity & Conservation* ISSN: (Online) (2014). 2311-9284, (Print) 0006-8241.

[2] Gmaraldeen, S. M., Magzoub, A. A., Badri, A. M., & Garbi, M. I. Antibacterial activity of *Acacia nilotica* fruits extract against pathogenic bacteria, *International Journal of Applied Research* (2016); 2(6): 103-106 ISSN *2*(June), 103-106.

[3] Ojewole, J. A. O., Mawoza, T., Chiwororo, W. D. H., & Owira, P. M. O. Sclerocarya birrea (A.Rich) Hochst. ['Marula'] (Anacardiaceae): A Review of its Phytochemistry , Pharmacology and Toxicology and its Ethnomedicinal Uses. *MDPI Molecules* (2018). https://doi.org/10.1002/ptr.3080

[4] Louis, H., Linus, M. N., Israt, A., Innocent, J., Amos, P. I., & Magu, T. O. Antimicrobial Activity of Stem , Leave and Root Plant Extract of *Sclerocarya birrea* and *Sterculia setigera* against Some Selected Microorganisms, Available online at www.worldscientificnews.com WSN 9 2. (2018). WSN 92(2) (2018) 309-326 (January), Mariod, A. A., & Abdelwahab, S. I. *Sclerocarya birrea :* Biochemical Composition, Nutritional and Medicinal Uses – *Natural Products: Research Reviews* (2012). *Vol. 1A 1.*

[5] Osawaru, M.E., Ogwu, M.C. and Ahana, C.M. Current Status of Plant Diversity and Conservation in Nigeria. Fourth National Biodiversity Report Abuja. (2010).79pp.

[6] Borokini, T.I. (2014). A systematic compilation of endemic flora in Nigeria for conservation management. *Journal of Threatened Taxa* 6(11): 6406-6426; http://dx.doi.org/10.11609/jott.o4010.6406-26

[7] Sahar Traoré Mohamed, Aliou Baldé Mamdou, Camara Aïssata, Cos Paul, Maes Louis, Pieters Luc.- The malaria co- infection challenge : an investigation into the antimicrobial activity of selected Guinean medicinal plants *Journal of ethnopharmacology* - ISSN 0378-8741-174(2015), p. 576-581 Full text (Publishers DOI): http://dx.doi.org/doi:10.1016/j.jep.2015.03.008

[8] Oni, P. I. (2001). Appendix 8, Forest Genetic Resources Priorities. Africa Report of the Fifth Session of the Panel of Experts on Forest Genetic Resources. Rome, Italy 8-11 December 2001, pp 52-100. FAO Rome. (http://www.fao.org/forestry/Forestry.asp).

[9] Varmola, M. and A. Del Lungo (2003). FAO. State of the world's forests. Rome, Italy Federal Department of Forestry, 1983. Nigerian Forestry Statistics 1983. 30 p (http://www.fao.org/forestry).

[10] Ogunwusi, A.A. (2013). The Impact of Forest Policy and Land Tenure System on Bamboo Development in Nigeria Raw Materials Research and Development Council, Abuja. *Public Policy and Administration Research* www.iiste.org ISSN 2224-5731(Paper) ISSN 2225-0972(Online). Vol.3, No.5, 2013 13.

[11] Yahaya Z. Ibrahim, Heiko Balzter, J€org Kaduk (2018), Land degradation continues despite greening in the Nigeria-Niger border region. *Global Ecology and Conservation journal* homepage: http://www.elsevier.com/locate/gecco.

[12] Charleston, D. S., Rieta Gols, Kees A Hordijk, Rami Kfir, Louise E M Vet, Marcel Dicke (2006). Impact

of botanical pesticides derived from *Melia azedarach* and *Azadirachta indica* plants on the emission of volatiles that attract Parasitoids of the diamondback moth to cabbage plants. *J Chem Ecol*., 2006 Feb; 32 (2):325-49.doi: 10.1007/s10886-005-9004-9.

[13] Bergdahl, Aaron D.; Hill, Alison, tech. coords. 2016. Diseases of trees in the Great Plains. Gen. Tech. Rep. RMRS-GTR-335. Fort Collins, CO: U.S. Department of Agriculture, Forest Service, Rocky Mountain Research Station. 229 p. www.fs.fed.us/rm/publications www.treesearch.fs.fed.us

[14] Umair M, Altaf M, Abbasi AM . An ethnobotanical survey of indigenous medicinal plants in Hafizabad district, Punjab-Pakistan. *PLoS ONE* (2017), 12(6): e0177912. https://doi.org/10.1371/journal.pone.017791

[15] Faruque, M. O., Uddin, S. B., Barlow, J. W., & Hu, S. (2018). Quantitative Ethnobotany of Medicinal Plants Used by Indigenous Communities in the Bandarban District of Bangladesh, *9*(February). https://doi.org/10.3389/fphar.2018.00040

[16] Ugwah, M. O., Etuk, E. U., Bello, S. O., Aliero, A. A., & Chinenye, J. (2013). Comparative studies of anti-ulcerogenic activities of three Nigerian medicinal plants: A preliminary evaluation. *Journal of Medicinal Plants Research*, *7*(9), 490-495. https://doi.org/10.5897/JMPR12.1018

[17] Pirker, H., Haselmair, R., Kuhn, E., Schunko, C., & Vogl, C. R. (2012). Transformation of traditional knowledge of medicinal plants : the case of Tyroleans (Austria) who migrated to Australia , Brazil and Peru.. *Journal of Ethnobiology and Ethnomedicine* (2012), 8:44 http://www.ethnobiomed.com/content/8/1/44

[18] Anka, Z. M., Singh, V., Gimba, S. N., & Singh, G. Antitoxic, Antifungal and Phytochemical Analysis of Medicinal Compounds of *Guiera senegalensis* Leaves, (2020). *10*(2), 148-152.

[19] Ekong, D.E.U., H. Herren, M.F. Ivbijaro (1992). Neem: A Tree for Solving Global Problems. Report of an Ad Hoc Panel of the Board of Science and Technology for International Development, National Research Council. ISBN: 0-309-58436-1, 152pp.

[20] Jones, E. W. 1955. Ecological studies of the rain forest of southern Nigeria IV. The plateau forest of the Okomu Forest Reserve, Part 1. The environment, the vegetation types of the forest, and the horizontal distribution of species. *Journal of Ecology* 43: 564-594.

[21] National Research Council. 1992. Neem: A Tree for Solving Global Problems. Washington, D.C., *National Academy* Press. 141 pp.

[22] Oates, J. F. 1995. The dangers of conservation by rural development – a case-study from the forests of Nigeria. Oryx 29: 115-122. Published online by Cambridge University Press: 24 April 2009. DOI: https://doi.org/10.1017/S0030605300020986.

[23] IUCN (2020). United Nations Environmental Program and World Conservation Monitoring Centre, International Union for Conservation of Nature, Red List of Threatened Species. IUCN 2020. *The IUCN Red List of Threatened Species*. Version (2020), 2. <https://www.iucnredlist.org> ISSN 2307-8235© *International Union for Conservation of Nature and Natural Resources*

[24] Nagesh VSDNA S, Muniappan M, Kannan I, Viswanathan S. Phytochemical analysis and docking study of compounds present in a polyherbal preparation used in the treatment of dermatophytosis. *Curr Med Mycol*. 2017; 3(4): 6-14. DOI: 10.29252/cmm.3.4.6

[25] Jones, E. W. 1956. Ecological studies of the rain forest of southern Nigeria IV. The plateau forest of the Okomu Forest Reserve, Part 2. The reproduction and history of the forest. *Journal of Ecology* 44:83-117.

[26] Loveland, T.R., B.C. Reed, J.F. Brown, D.O. Ohlen, Z. Zhu, L. Yang, and J.W. Merchant. 2000. Development of a global land cover characteristics database and IGBP discover from 1 km AVHRR data. *International Journal of Remote Sensing.* 21:1303-1330.

Chapter 5

Endemic Vascular Plants from the Coromandel Coast of Tamil Nadu, Southern India

Dhatchanamoorthy Narayanasamy and Balachandran Natesan

Abstract

Intensive botanical survey was done more than two decades on Tropical Dry Evergreen Forest from four coastal districts *viz*. Cuddalore, Kancheepurm, Nagai and Villupuram of Tamil Nadu and Union Territory of Pondicherry. A total of 87 protected (hillocks, reserve forest) and unprotected (sacred groves, unclassified vegetation) sites were regularly studied from five districts. From this study 82 endemic taxa were enumerated. In addition 25 endemic species were added for the analysis through literature screening and herbarium consultation from 10 Coromandel coastal districts of Tamil Nadu. In all 107 species were recorded, among them 19 are trees, 18 shrubs, 9 climbers and 61 herbs. Distribution of these endemic species were analysed and categorised into endemic to the country, peninsular India, southern India, Eastern and Western Ghats, state and district level. Interestingly the study found that some endemic species were disjunctly distributed in between districts, states, ghats, climatic regimes and bioregions. The disturbance, threat status and conservation measures of few endemic and IUCN red listed species were also studied and discussed.

Keywords: conservation, Coromandel Coast, disjunction, endemism, Pondicherry, Tamil Nadu, IUCN status, tropical dry Evergreen Forest

1. Introduction

India is one of the 18 extremely diverse and top 10 species-rich countries of the world, in which a total of 4381 taxa belonging to 1007 genera and 176 families, including 4303 angiosperms, 12 gymnosperms and 66 pteridophytes, out of 18,043 species have been confirmed as endemic to India [1]. Recently, in angiosperms a total of 58 genera have been identified as endemic to India, of which 49 are confined to Peninsular India [2]. Though the term 'endemism' coined during eighteenth century Chatterjee [3, 4] was the first who studied the endemism of the Indian flora. He was considered 6850 species that are unique to this region (61% of flowering plants), of which 3169 species are restricted to Himalayas and 2045 to Peninsular India (PI). Blasco [5] was estimated about 1268 endemic dicotyledons to south India; however Nayar [6] recorded 2100 flowering plants endemic to PI. Later, Nayar [7] reported 141 genera endemic to India; while Ahmedullah & Nayar

[7] found 55 genera endemic to PI of which 45 are monotypic [7]. Recently, Irwin & Narasimhan [9] enumerated only 49 genera which are endemics to India, excluding several genera based on nomenclatural changes and extended distribution. Nayar [10] categorised the endemic genera of India into 3 patterns based on the distribution *viz*. Himalayan Endemic Genera, Peninsular Indian Endemic genera and Andaman & Nicobar Islands Endemic genera. Meanwhile there are no families which are endemic to India [8].

In India, Western Ghats has much more endemic (2116 species) taxa than rest of India. State-wise analysis Tamil Nadu ranking first with 410 species, followed by Kerala (357), Maharashtra (278) and Andaman and Nicobar Islands together contributes with 278 taxa [2]. In Eastern Ghats (EG), a total of 166 endemic taxa, under 117 genera and 43 families are known to occur, of which 129 dicots and 46 are monocots. Sudhakar Reddy & Raju [11] recorded 400 endemic spermatophytes from the EGs of Andhra Pradesh and their adjacent coastal plains.

The Flora of Tamil Nadu published during 1983, 1987 and 1989 in 3 volumes, and the report after Betty & Ramachandran [12] was added 192 taxa belonging to 130 genera and 61 families between the period 1989 and 2013. These additions were compiled from research articles, unpublished thesis and research reports by several botanists [13–19]. Among them 87 taxa are new to the science as well as endemic to the state of Tamil Nadu. The high concentration of endemic plants once again proves that the southern India is one of the top ten mega biodiversity hotspot area. It directly reflects the habitat stability, environmental quality, and rich biodiversity and conservation values in a specific area. However the Coromandel Coast is another unique bioregion, as flood plain and a buffer zone between the hill range of Eastern Ghats and Bay of Bengal; and this was not studied or updated since Roxburgh [20]. The main aim of this work is to explore the wealth and threat status of endemic plants diversity from the Coromandel Coast and especially from the fragile ecosystem of Tropical Dry Evergreen Forests (TDEF), was classified by Champion and Seth [21].

2. Materials and methods

The geographical area of Tamil Nadu is 130,058 km^2 and has roughly rhomboidal shape in appearance. It lies between 8° 5′-13° 35’ N latitude and 76° 15′-80° 20′E longitude. The state occupies 4.08% of the total area of the country. It has the coast line of 990 km at east and land boundary of 1200 km towards west. The state is divided into 38 districts of which 13 districts lies on the east coast. The natural land mass of the state was divided into the Eastern Ghats, Coastal Plains, Central plateau and Western Ghats. Most part of the 13 coastal districts considered as the Coromandel Coast of Tamil Nadu (**Figure 1**). The entire coast of Tamil Nadu, is chiefly sandy with outcrops of rocky headlands at Kancheepuram, Kanyakumari, Tirunelveli and Villupuram districts. The coastal vegetation had further subdivided into Strand, Estuarine and Coastal Tropical Dry Evergreen Forest types (Nair and Henry [22]).

2.1 Study area

The present study has included 13 coastal districts of Tamil Nadu; Karaikal and Puducherry regions from Union Territory (UT) of Pondicherry along the Coromandel Coast of south India. However, regular intensive survey was done since 1996 to till date on four coastal districts (Cuddalore, Kancheepuram, Nagai and Villupuram) and two regions (Puducherry and Karaikal) from the UT of Pondicherry. Forest cover of Cuddalore is 444 km^2 (11.98%) out of 3706 km^2, Kancheepuram 372 km^2 (8.31%) out of 4474 km^2 and Villupuram 1011 km^2

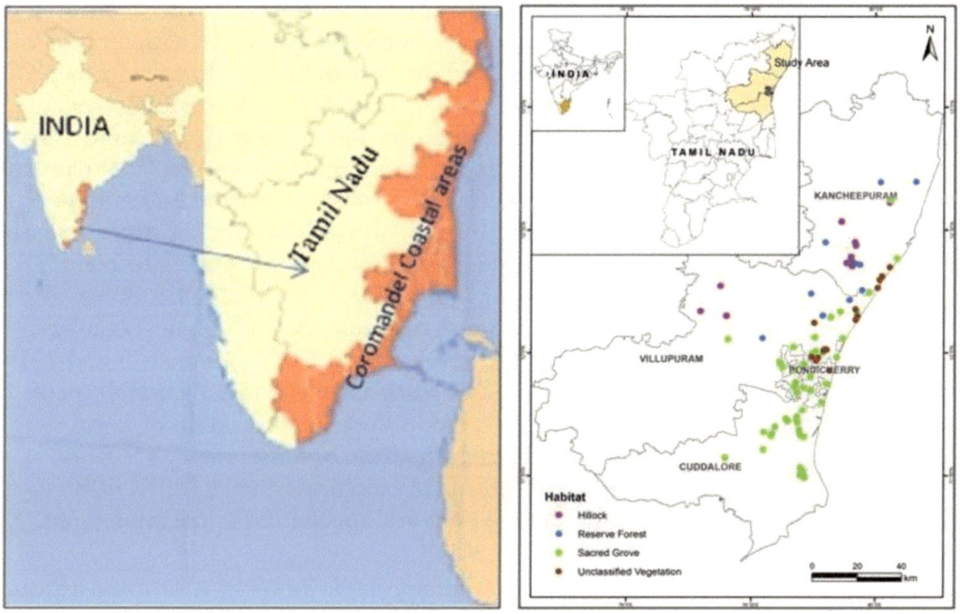

Figure 1.
Study area - Coromandel Coast and detailed study at district level.

(14.06%) out of 7190 km^2 geographical area [23]. The forest cover at Union Territory of Pondicherry is 50.06 km^2 (10.43%) out of 480 km^2 area [24].

Geologically, part of Cuddalore and Villupuram districts belonged to the formation of Cuddalore sandstone during Miocene period [25]. The soil along the coast is sandy loam or red ferralitic and in certain places covered with alluvial deposits and becoming clayey at interior [26, 27]. The coastal plains are extending up to 40–60 km [28] and are overlained by a thin soil layer supporting agriculture. The substratum erupted into hillocks and mounds at Kancheepuram and Villupuram districts and into undulating terrain in Cuddalore districts. The scattered hillocks rise up to 450 m with interrupted vegetation among the charkonite or gneiss rocks. The natural vegetation is mostly found on less fertile and red ferralitic soil, whereas black clay and alluvial soils were brought under cultivation [29].

A typical maritime tropical climate with dissymmetric rainfall regime occurs in the study area. The mean annual rainfall recorded during 2007–2016 period was 1256 mm with mean rainy days of 56 per year. The minimum temperature 17.7°C is in January, maximum temperature 40.5°C in May and the mean is 28.5°C. The average relative humidity is 76% and the weather is generally cool during December to January with the late nights dewy. Dry weather prevails during April to June. Wind speed ranges from 5 to 9 km/h during July to September but extremely higher during the cyclonic days, during October to December [30].

2.2 Field survey

Four types of vegetation covers including micro and macro habitats *viz.*, Hillocks (HL), Reserve Forest (RF), Sacred Groves (SG) and Unclassified Vegetation (UC) were identified using Geological Survey of India (GSI) map, Google map and interview with people. The areas of HL vegetation are ranging from 680 to 2200, RFs from 100 to 350, SGs from one to 40 and UCs from 0.5 to 35 hectares (ha). The elevation of HL was found between 150 and 450, RFs 100 to 350, SGs sea level to 80 and UC sea level to 40 m a.s.l.

Botanical surveys were made extensively once in a week on 87 sites from five districts with a team of four members, visiting each and every site with an interval of 4–6 months and monitored pre-monsoon and post-monsoon changes from 1996 to 2019. These sites were geo-referenced with Garmin Global Positioning System (GPS), followed by intensive species enumerations including herbs, shrubs, trees and climbers; collection of voucher samples and photographed the key characters of the plants. A total of 15,316 herbarium sheets were prepared from the sample collections and deposited at AURO! Herbarium, Auroville, India. The nomenclature of all plant species recorded in this study was verified in www.plantsoftheworld online.org. In addition, Endemic plants of Indian Region (Ahmedullah & Nayar [10]), The Flora of the Gulf of Mannar, Southern India [31], Endemic Vascular Plants of India [2], Plant Discoveries [32], research articles between 2013 and 2019 period and different herbarium such as Saint Joseph College (SJC) Tiruchirapalli, Madras Herbarium (MH!) Coimbatore, Foundation of Revitalization of Local HealthTraditions (FRLH) Bangalore, French Institute of Pondicherry (HIFP) Puducherry, and Sri ParamaKalyani Center for Ecological Studies (SPKCESH) Tirunelveli was referred and enriched the endemic species list to the study area.

2.3 Analyses

Based on phytogeographical distribution six groups of endemic regions were categorised, such as 1. the state Tamil Nadu, 2. Eastern Ghats (EG), 3. EG & Western Ghats (WG), 4. Southern India (SI), 5. Peninsular India (PI) and 6. Entire India except Himalayas. In addition, disjunct nature of distribution of these endemic species between or among the regions was also studied. Site disturbances such as browsing, cutting, lopping, and clear felling, encroachment for cultivation purposes, construction of big modern temple, construction and widening the metal road, digging irrigation channels and cementing the thrashing floor were studied and categorised into low, medium and high by following Venkateswaran & Parthasarathy [33]. High ranks signify high levels of anthropogenic disturbance in the forests. The disturbance codes were co-related with four life forms and threat status of all species was verified with www.iucnredlist.org (version 2020–2). Threat assessment and possible conservation measures were undertaken on few endemic and endangered species through Auroville greening and Botanical Garden Projects.

3. Results

Through our regular field study on 87 sites, 25 (SG) sites are from Cuddalore, 22 (8 HL, 6 RF, 4 SG, 4 UC) from Kancheepuram, one site (1RF) from Nagai, 28 (3 HL, 4 RF, 13 SG, 8 UC) from Villupuram districts of Tamil Nadu, 10 (8 SG, 2 UC) from Puducherry region and entire Karaikal region, Union Territory of Pondicherry. Altogether, 1197 species were listed from 127 families and 584 genera, of which 196 species are trees, 113 shrubs, 172 climbers and 716 herbs. Through literature screening and referring the herbarium 25 endemics were added. Finally a total of 107 endemic species were compiled for the Coromandel Coast of Peninsular India and analysed. Of which 19 species are trees, 18 shrubs, 9 climbers and 61 herbs (**Figure 2**). These endemic species were represented by 74 genera and 33 families, of which Leguminosae (17 species from 10 genera) is the dominant family followed by Acanthaceae (13 species from 5 genera), Euphorbiaceae (10 species from 5 genera) and Poaceae (9 from 9 genera). The other dominant families are Apocynaceae and Rubiaceae, had 4 species each (**Figure 3**).

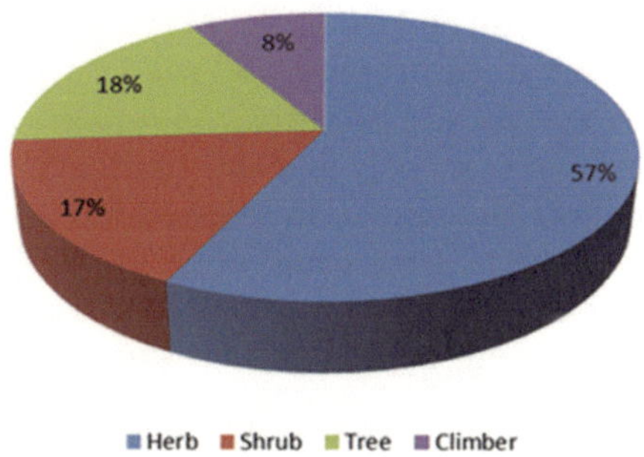

Figure 2.
Endemic plants of CC and their habits representation.

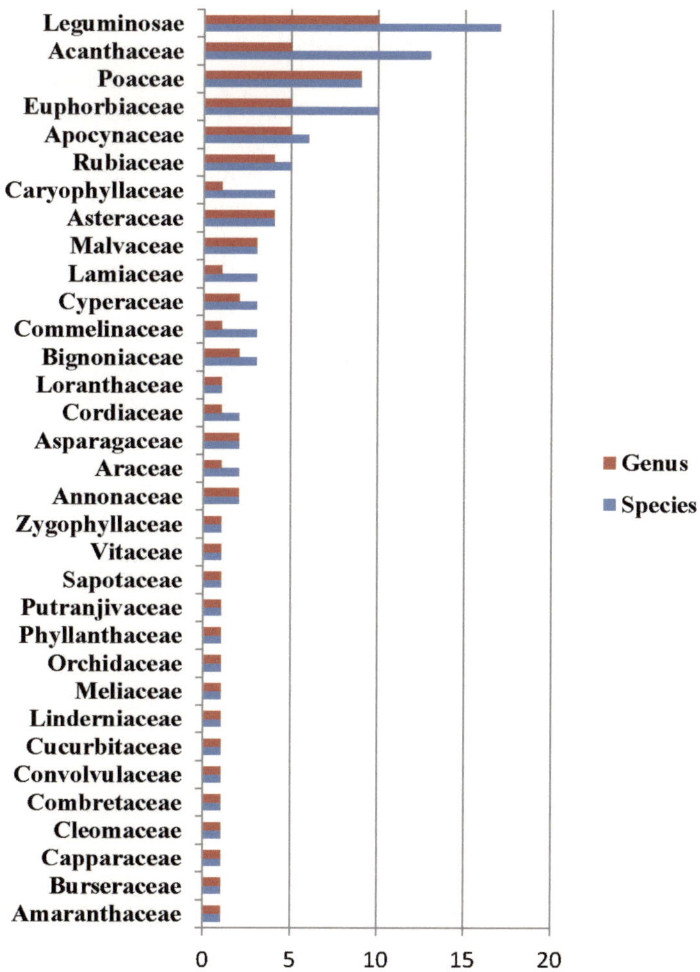

Figure 3.
Detail of endemic plant families, genera and species.

3.1 Disturbance

Qualitatively, classified disturbances were noted from four vegetation types (**Table 1**; **Figure 4**). These disturbance codes were correlated with life form, threat status and ethno-botanical values of endemic species. In general, from 50 to 65% of species were represented in disturbed category. Maximum numbers of species are encountered at medium level of disturbance, followed by undisturbed, low and high level of disturbance. Reasons for the threat are: root of *Decalepis hamiltonii* and fruit of *Phyllanthus indo-fischeri* were extensively collected for their medicinal properties; *Justicia beddomei*, *Leucas wightiana* and *Lindernia minima* are endangered due to narrow and disjunct distribution; *Derris ovalifolia*, *Drypetes porteri*, and *Koilodepas calycinum* are facing habitat loss; *Pterocarpus santalinus* was threatened due to illegal logging; *Sterculia populifolia* living with rocky outskirts; and the herbs are generally facing pressure by grazing.

3.2 Distribution

For an understanding based on geographical distribution the 107 endemic species that recorded from the CC plains were divided into eight groups and their representations are: 1. India (excluding Himalayas & north east) (6 species), 2. Peninsular India (22 species), 3. Southern India (28 species), 4. Southern India with one or two states of north India (11species), 5. Eastern and Western Ghats (5 species), 6. Eastern Ghats (5 species), 7. Tamil Nadu (27 species) and 8. Dispersed in different states (3 species) (**Figure 5**). The distribution of six species across the country is *Crotalaria pusilla*, *Dolichandrone falcata* and *Hardwickia binata, Iseilema anthephoroides*, *Lophopogon tridentatus* and *Scleria stocksiana* (**Figure 6**). *Crotalaria willdenowiana*, *Deccania pubescens* var. *candolleana*, *Polycarpaea corymbosa*, *Pterocarpus santalinus* and *Sterculia populifolia* were spotted only in EGs of Andhra Pradesh, Karnataka and Tamil Nadu (**Figure 7**). Five species *viz*. *Derris ovalifolia*, *Mallotus resinosus* var. *muricatus*, *Mussaenda glabrata, Rhynchosia courtallensis* and *Tetrastigma tamilnadense* was found both in EG and WG (**Figure 7**). At regional

Details		Undisturbed	Disturbed		
			Low	Medium	High
Life form	Herbs (61)	16	12	24	9
	Shrubs (18)	5	6	3	4
	Trees (19)	7	3	4	5
	Climbers (9)	4	1	3	1
Threat status	IUCN categorised (7)	3	1	2	1
	From publications (10)	3	3	3	1
	Not evaluated (90)	27	19	29	15

Table 1.
Correlation of endemic plants life form and threat status with disturbance gradience.

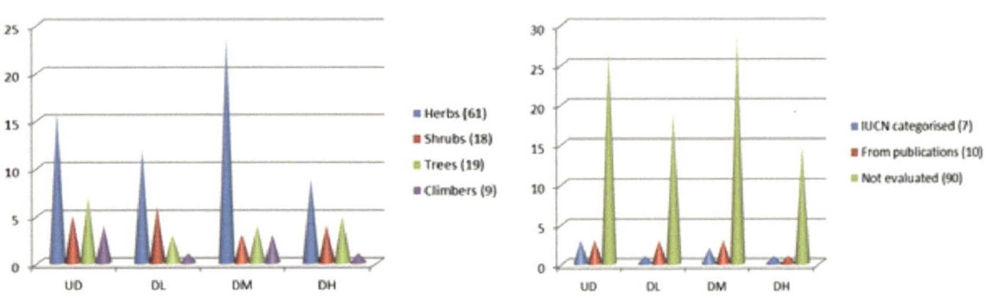

Figure 4.
Disturbance index with life-form and threat status.

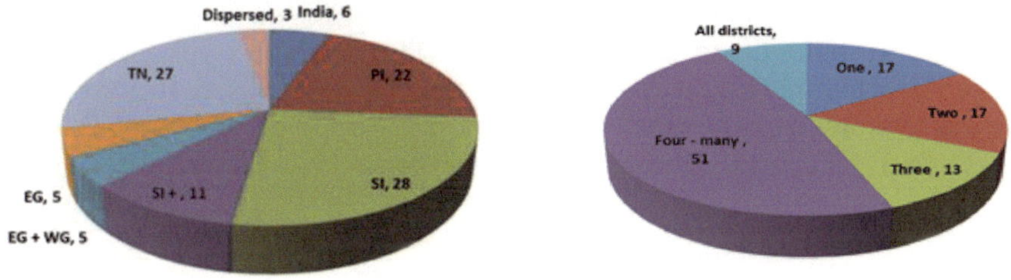

Figure 5.
Geographical representation of endemics at regional and district level of Tamil Nadu.

Figure 6.
Representation of endemic plants distributed at national and regional level.

level south India has 28 species followed by 27 species restricted to Tamil Nadu and 22 taxa represented from peninsular India (**Figure 5**).

The representations of species at different districts of Tamil Nadu are analysed, and the study shows that the species found in only one (17 species), two (17 species), three (13), four – many (51 species) and in all districts (9) were recorded (**Figure 5**).

3.3 Narrow or steno endemics

Thirteen species were showed very narrow distribution, found in only one district. They are *Acrachne henrardiana* of Poaceae (Pudukkottai district), *Barleria*

Figure 7.
Representation of endemic plants distributed at state level and RET endemics.

durairajii of Acanthaceae (Thoothukudi), *Caralluma adscendens* var. *gracilis* of Apocynaceae (Pudukkottai), *Cordia ramanujamii* of Cordiaceae (Villupuram), *Derris gamblei* of Leguminosae (Pudukkottai), *Huberantha senjiana* of Annonaceae (Villupuram), *Jatropha villosa var. ramnadensis* of Euphorbiaceae (Ramanathapuram), *Lepidagathis pungens* of Acanthaceae (Tirunelveli), *Leucas anandaraoana* of Lamiaceae (Ramanathapuram), *Polycarpaea diffusa* and *Polycarpaea majumdariana* of Caryophyllaceae (Thoothukudi and Tirunelveli), *Sporobolus hajrae* of Poaceae (Pudukkottai) and *Theriophonum sivaganganum* of Araceae (Ramanathapuram); of which C. ramanujamii and H. senjiana are recently described neo-endemics (**Figure 8**).

3.4 Disjunct distribution

The analysis shows interesting disjunctions between: 1. The districts of Tamil Nadu, 2. SI and north-east and 3. SI and north-west, and 4. SI and trans-Himalaya (**Figure 9**). *Garnotia elata* of Poaceae and *Glossocardia bosvallia* of Asteraceae are sharing their region between SI and Uttar Pradesh; *Indigofera mysorensis* of Leguminosae and *Jatropha tanjorensis* of Euphorbiaceae between SI and West Bengal; *Manisuris myurus* of Poaceae between SI & Manipur; *Oldenlandia attenuata* of Rubiaceae between EG and Uttar Pradesh; *Leucas diffusa* between SI and Delhi;

Figure 8.
Recently described endemic species.

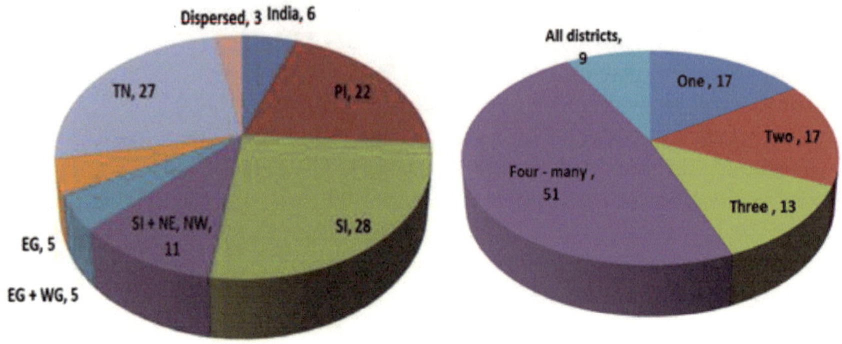

Figure 9.
Representation of state and district level distribution of endemics.

L. wightiana and *Senna montana* between SI and Gujarat; and *Tricholepis radicans* between SI and Rajasthan are some important disjunct distributions between two major geographical regions (**Figure 10**).

In Tamil Nadu at district level analysis (**Figure 9**) found that *Blumea eriantha* of Asteraceae (Coimbatore-Villupuram), *Chlorophytum malabaricum* of Asparagaceae (Nilgiri-Villupuram), *Cordia diffusa* of Cordiaceae (Coimbatore-Kancheepuram), *Drypetes porteri* of Euphorbiaceae (Theni-Villupuram), *Justicia beddomei* of Acanthaceae (Tirunelveli-Villupuram), *Melothria angulata* of Cucurbitaceae (Dindigul-Villupuram), *Mussaenda tomentosa* of Rubiaceae (Tirunelveli-Villupuram), *Rhynchosia courtallensis* of Leguminosae (Tirunelveli-Villupuram), *Tetrastigma tamilnadense* of Vitaceae (Tirunelveli-Villupuram) and *Trachys narasimhanii* of Poaceae (Chennai-Ramanathapuram) are showing great disjunction between the north and south district (**Figure 10**).

3.5 Doubtful endemics

There are six species *viz*. *Ceropegia mannarana* of Apocynaceae, *Leucas nepetifolia* of Lamiaceae, *Mariscus clarkii* of Cyperaceae, *Stenosiphonium parviflorum* of Acanthaceae *Sehima sulcatum* and *Zenkeria elegans* of Poaceae are considered as doubtful endemics because Singh et al. [2], Krishnamurthy et al. [34], Henry et al. [35] and Ahmedullah & Nayar [10] treated them as endemic whereas the www.plantsoftheworldonline.org for the former four species and Kabeer & Nair [13] for the latter two Poaceae members had marked them as non-endemic. Meanwhile, Krishnamurthy et al. [34] was included the two latter species under "endangered" list.

3.6 Threat status

According to IUCN [36], a total of 90 (84.11%) out of 107 endemic species were listed as "not evaluated" and their population status in the wild habitats is also unknown. So far, only seven species *viz*., *Aglaia elaeagnoidea* (LC [37]), *Decalepis hamiltonii* (EN-A2cd, [38]), *Drypetes porteri* (EN-B1 + 2c, [39]), Koilodepas calycinum (EN B1 + 2c, [39]), *Lindernia minima* (EN-B1ab + 2ab, [40]), *Pterocarpus santalinus* (EN-B1 + 2de, [41]), and *Sterculia populifolia* (CRD, [39]) are assessed according to IUCN criteria (**Figure 7**). From the published information found that 10 species such as, *Chrysopogon verticillatus* [13], *Crotalaria willdenowiana* [34], *Derris ovalifolia* [34, 42, 43], *Dipcadi montanum* var. *madrasicum* [10, 44], *Discospermum sphaerocarpum* [45], *Justicia beddomei* [10], *Leucas wightiana* [10],

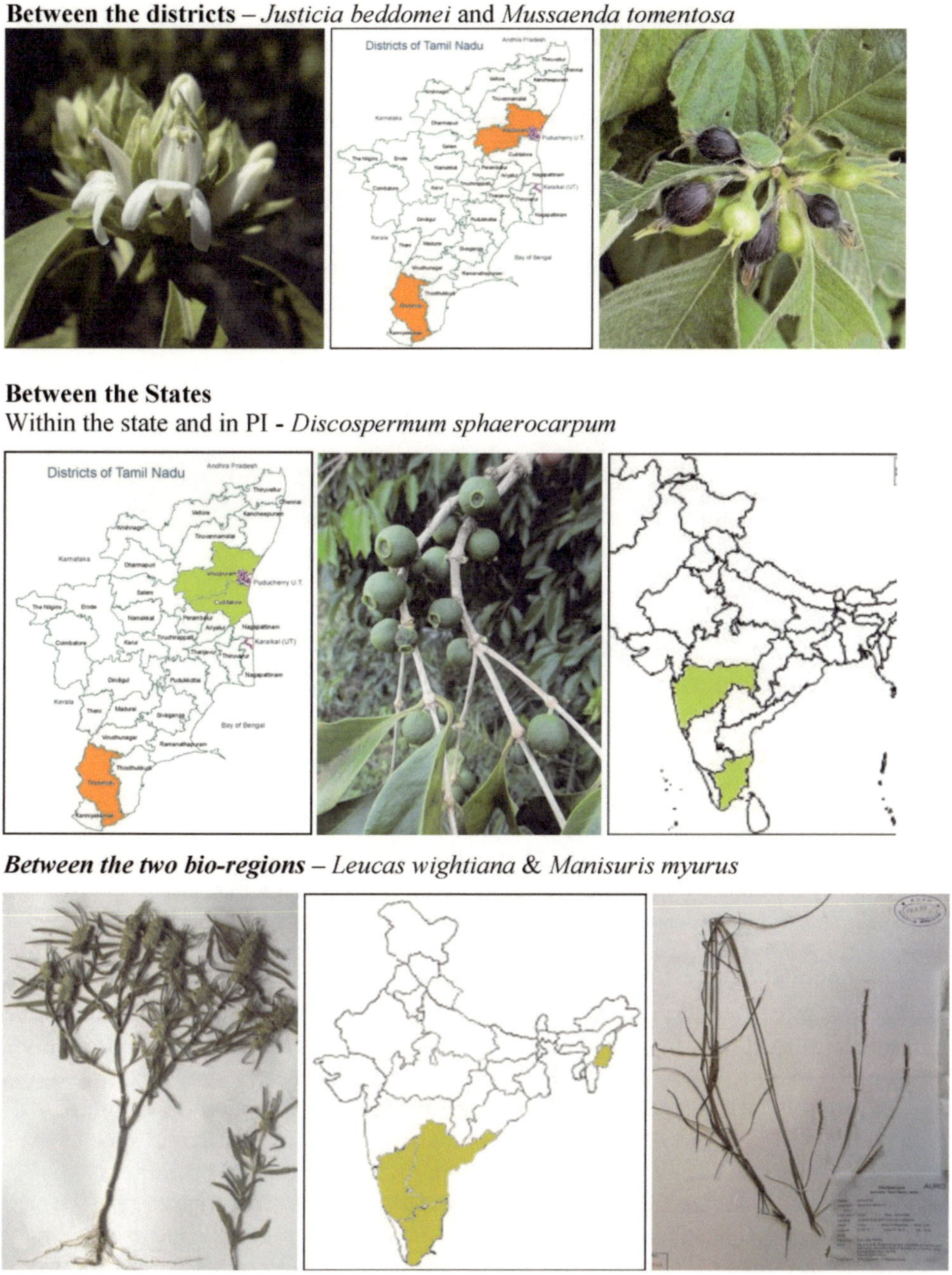

Figure 10.
Disjunct distribution of endemics.

Melothria angulata [22], *Rhynchosia courtallensis* [34] and *Sarcostemma intermedium* [34] were came to known as rare or endangered or threatened endemic species.

3.7 Threat assessment and conservation measures

Rapid Assessment Workshop on Conservation of Tropical Dry Evergreen Forest was conducted by Auroville Green Group, Auroville in collaboration with Foundation for Revitalization of Local Health Traditions, Bengaluru held between 5 and 7 March, 2002. A team of 32 field botanist and experts involved and assessed 48

Figure 11.
Conservation status of Hildegardia populifolia *and its habitat.*

species but not published, in which 11 species such as *Acrachne henrandiana*, *Aglaia elaeagnoidea, Cadaba trifoliata*, *Caralluma indica*, *Carissa salicina*, *Ceropegia juncea*, *Derris ovalifolia*, *Discospemum sphaerocarpum* (syn. *Tricalysia sphaerocarpa*), *Habenaria roxburghii*, *Manisuris myurus*, and *Pterospermum xylocarpum* are endemic.

Since 1996, extensive effort was employed by Auroville Green Group to conserve on Tropical Dry Evergreen Forest, is an endangered forest type in India. More than one lakh seedlings from 250 native species produced every year planted and developed 'Green Ring' in 2000 hectares at Auroville. Basically they attempted to conserve all the TDEF species; in addition they were concentrated on endemic, endangered and rare Indo-Sri Lankan elements too. Few such endemic species are *Commiphora berryii*, *Deccania pubescens, Derris ovalifolia*, *Discospermum sphaerocarpum, Dolichandrone atrovirens, Dolichandrone falcata*, *Drypetes porteri, Hardwickia binata, Justicia beddomei, Manilkara roxburghiana, Miliusa eriocarpa, Pterocarpus santalinus*, *Pterospermum xylocarpum, Sterculia populifolia* and so on. They are standardised the germination through different techniques for most species but they are not successful in *Terminalia paniculata* of Combretaceae and two non-endemic native species *viz*. *Anogeissus latifolia* of Combretaceae and *Hugonia mystax* of Linaceae.

3.7.1 Sterculia populifolia, *threat assessment: a status survey*

The conservation status of *Hildegardia populifolia* (syn of *Sterculia populifolia*) was kept in different category by the experts of India. This species was assessed first as 'Critically Endangered' by The World Conservation Monitoring Center WCMC [13] (**Figure 11**) and there are no update to till date. Literature screening showed with other categories as viz. 'Endangered' by Krishnamurthy et al. [34], Walter and

Gillet [46]; 'Vulnerable' by Reddy et al. [47]; 'Very Rare' by Rao and Pullaiah [48]. Whereas Sarcar and Sarcar [49] studied the population, propagation and conservation of the species and recommended the species as critically endangered. Madhavachetty et al. [50] was reported this species occurrence from Ankalamma Konda in Chittoor district. Recently Jaikrishnan [51] was made intensive population survey, counted all mature individual and assessed the new recruits at Gingee hills of Villupuram District, Tamil Nadu and finally concluded this species as 'critically endangered'. The taxa which are under threatened status since 1997 categorised by IUCN and Botanical Survey of India was published Red data books in three volumes during 1987–1990, in which *Hildegardia populifolia* was also treated as endangered.

4. Discussion

The floral diversity in any state or country or world the dominant families are Leguminosae, Poaceae, Orchidaceae, Acanthaceae, Euphorbiaceae, Asteraceae, Apocynaceae, and Rubiaceae so on. This present endemic study was more intrinsically matching with, regional [52] and national [1, 2] studies/analysis. Also, the study reports from the coastal plains [26, 53–58] favoured the present study. Out of 107 endemics 27 species are restricted to the state Tamil Nadu, 5 to EG, 5 to EG & WG, 28 to SI, 11 species sharing between SI with different states of north India, 22 to PI, to entire India, and 3 to elsewhere. These data were enlightened the richness of plant diversity at the Coromandel coast, especially from the TDEF of Tamil Nadu and supporting the 'coastal zone' as one of the endemic centres of India.

The International Union for Conservation of Nature and Natural Resources [46] assessed the global threat status of 33,418 species of Angiosperms, of which 1215 species are reported from India, of which 690 (55.8%) species were evaluated as Indeterminate (I). From this study 80% of endemic species were in 'not evaluated' category. According to Isik [59] three-quarters of narrow endemic species of plants and animals are known to have become extinct due to habitat loss or fragmentation. So as Nair [60] statement "it is very essential that rare, threatened and presumed extinct taxa should be repeatedly searched for in their type localities", should be strictly followed and need to do their population assessment status from time to time. Narrow range and regional level assessments are making ambiguous with the IUCN category, so the WCMC should follow these publications and update them to the relevant species.

Majority of endemic species are isolated due to geographical, ecological, edaphic and climatic barriers and these fragmented patches of vegetation were more pronounced in EG for the point of conservation [61, 62]. This condition was more privileged to the narrow endemic species like *Cordia ramanujamii*, *Huberantha senjiana* and *Mussaenda tomentosa* at Pakkam Malai reserve forest; Gingee hills of EG [53]. Meanwhile, the disjunctly distributed species like *Blumea eriantha*, *Cyanotis tuberosa*, *Drypetes porteri*, *Derris ovalifolia*, *Discospermum sphaerocarpum*, *Indigofera mysorensis*, *Justicia beddomei, Leucas wightiana*, *Manisuris myuros*, *Melothria angulata* between the two Ghats/bioregion have to be considered as crucial for the conservation actions.

It was estimated that 2–25% of plant species will become extinct or committed to extinction in tropical forest approximately in next years [63]. It is also opined that 22–47% of species might have already become threatened [64]. In India TDEF occupies about 2482.52 km^2 (1.61% of the country territory), in which Tamil Nadu has only 41.08 km^2 (0.1%) [65]. According to Krishnamurthy et al. [34] the TDEF found along the Coromandel Coast is an 'endangered forest' type. The degree of threat and richness of endemism is one of the major aspects in prioritising the areas for conservation. In this paradigm, Jain & Rao [66] statement "if endemic species are eliminated from our country it will mean that they will be annihilated from the whole world, will be loss to science, will be struck off the roles of biological resources of this earth" should be profoundly considered. In all, highly fragmented form of TDEF ecosystem, indeterminate IUCN status of narrow endemics and their disjunct distribution with different bioregions of India should be considered as high priority for the assessment and conservation programs at national, regional and state level in regular intervals.

Acknowledgements

The authors are thankful to Walter Gastmans, curator of AURO herbarium and Dr. Raphael Mathevet, Head of Ecology, French Institute of Pondicherry for their constant support and encouragement; financial support for the survey from 1994 to 2000 by Danish Government through Foundation for Revitalization of Local Health Traditions, Bengaluru; 2001-2004 by European Commission; 2005-2010 by Auroville Coastal Development Centre; WCT-Small Grant 2018-2019 Phase I; Head of Head of Forest Force, Chennai and District Forest Officer, Villupuram, Tamil Nadu granted permission to do the botanical survey in RFs; Mr. Paul Blanch Flower and Mr. Jaap Hollander accompanied all the times during the survey and provided photos of some species.

Author details

Dhatchanamoorthy Narayanasamy[1*] and Balachandran Natesan[2]

1 School of Conservation of Natural Resources, The University of Trans Disciplinary Health Sciences and Technology (TDU), Yelahanka, Bengaluru, India

2 Ecology Department, French Institute of Pondicherry, Puducherry, India

*Address all correspondence to: dhatcha@gmail.com

References

[1] Singh P, Dash SS. Plant Discoveries 2013. Botanical Survey of India, Kolkata; 2014

[2] Singh P, Karthigeyan K, Lakshminarasimhan P, Dash SS. Endemic Vascular Plants of India. Botanical Survey of India, Kolkata; i-ixvi:1-339 pp

[3] Chatterjee D. Studies on the endemic flora of India and Burma. J. Royal Asiatic Soc. Bengal. 1939 (1940). **5**:19-69

[4] Chatterjee D. Floristic pattern in Indian Vegetation. Proc. Summer School Botany, Darjeeling, New Delhi. 1962: 32-42

[5] Blasco F. 1971. Montagnes du sud del'nde: forets, savanes, ecologie. Inst, Fr. Pondicherry, Trv, Sec. Sci. tech. Tome vol. X 436 pp.

[6] Nayar MP. Changing pattern of distribution of endemic genera (angiosperm). Journal of Economic and Taxonomic Botany. 1977;**1**:99-110

[7] Nayar MP. Endemism and patterns of distribution of endemic genera (angiosperms) in India. J. Eco. Tax. Bot. Vol. 1980a;**1**:99-110

[8] Irwin SJ, Narasimhan D. Endemic genera of angiosperms in India: A review. Rheedea. 2011;**21**:87-105

[9] Nayar MP. Endemic flora of Peninsular India and its significance. Bull. Bot. Surv. India. 1980b (1982);22 (1-4):12-23

[10] Ahmedullah M, Nayar MP. Endemic Plants of Indian Region. Vol. 1 (Peninsular India). Botanical Survey of India; 1986. pp. 1-261

[11] Sudhakar Reddy C, Raju VS. Endemic spermatophytes of Andhra Pradesh, India. Proc. AP Akademi of Sciences. 2008;**12**:48-75

[12] Betty T, Ramachandran VS. Additions to the Flora of Tamil Nadu. J. Biod. Photon. 2014;**113**:355-359

[13] Kabeer KAA, Nair VJ. Flora of Tamil Nadu – Grasses. In: 1–525 pp. Botanical Survey of India, Coimbatore. 2009: 525 pp

[14] Manickam VS, Murugan C, Sundaresan V, Jothi GJ. Flora of Tirunelveli Hills: Southern Western Ghats. Bishen Singh Mahendra Pal Singh, Dehra Dun; 2008

[15] Matthew KM. Flora of the Palni Hills. Vol. 1- III. Tiruchirapalli: The Rapinat Herbarium; 1999

[16] Ramachandran VS, Balasubramaniam V, Pandikumar P. Additions to the grass flora of Tamil Nadu. Journal of Bombay Natural History Society. 2006;**102**(3):362-365

[17] Uma Maheshwari P, Daniel P. Flora of Gulf of Mannar. Kolkata: Botanical Survey of India; 2001. pp. 1-688

[18] Viswanathan MB, Manikandan U. A new species of *Hedyotis* (Rubiaceae) from India. Edinburgh Journal of Botany. 2008;**65**(3):387-392

[19] Yarrayya K, Murthy GVS, Ratna Kumar PK. Addition of sedges to the flora of Tamil Nadu state. Ind. J Plant Sci. 2015;**4**(4):57-59

[20] Roxburgh W. Plants of the Coast of Coromandel; Drawings and Descriptions. Vol. 1-3. London: Prn. W Bulmer & Co.; 1795-1805

[21] Champion HG, Seth SK. Revised Survey of the Forest Types of India. New Delhi: Manager of Publications; 1968

[22] Nair NC, Henry AN. Flora of Tamil Nadu, India. In: Series 1: Analysis vol 1. Botanical Survey of India: Coimbatore; 1983

[23] Annamalai R. Tamil Nadu Biodiversity Strategy Action Plan. Government of Tamil Nadu, Chennai: Tamil Nadu Forest Department; 2004 143 pp

[24] Forest Survey of India. Forest Cover In: India State Forest Report 2013, FSI, Dehra Dun, 11–32. http://fsi.nic.in/cover_2013/sfr_forest_cover.pdf accessed on 25 June 2020.

[25] Meher-Homji VM. Notes on some peculiar cases of Phytogeographic distribution. J. Bombay Nat. Hist. Society. 1970;**67**:81-87

[26] Meher-Homji VM. On the origin of tropical dry evergreen forest of South India. Int. J. Eco. Environ. Science. 1974a;**1**:19-39

[27] Meher-Homji VM. A new classification of the Phytogeographic zones of India. Indian J. Botany. 1984;7: 224-233

[28] Mani MS. Ecology and Biogeography in India. The Netherlands: BV Publishers; 1974

[29] Marlange M, Meher-Homji VM. Phytosociological studies in the Pondicherry region. J. Indian Bot. Society. 1965:167-182

[30] Meher-Homji VM. The climate of Cuddalore: A bioclimatic analysis. Geographical Review of India. 1974b;**36** (1):1-22

[31] Daniel P, Umamaheswari P. The Flora of the Gulf of Mannar, Southern India. Calcutta: Botanical Survey of India; 2001. p. 688

[32] Plant Discoveries. Botanical survey of India. Kolkatta. 2007-2018

[33] Venkateswaran R, Parthasarathy N. Tropical dry Evergreen forests on the Coromandel Coast of India: Structure. Composition and Human disturbance. Ecotropica. 2003;**9**:45-58

[34] Krishnamurthy K, Murugan R, Ravikumar K. Bioresources of the eastern Ghats, their conservation and management. Bishen Singh Mahendra Pal Singh: Dehra Dun; 2014. 824 pp

[35] Henry AN, Chitra V, Balakrishnan NP. Flora of Tamil Nadu, India Series 1: Analysis. Vol 3. Botanical Survey of India, Coimbatore. 1989.

[36] IUCN 2020. The IUCN Red List of Threatened Species. Version 2020-2. http://www.iucnredlist.org. Downloaded on 28th June 2020

[37] Pannell CM. Aglaia elaeagnoidea. The IUCN Red List of Threatened Species 1998: e.T33711A9804005. http://dx.doi.org/10.2305/IUCN.UK.1998.RLTS.T33711A9804005.en. Downloaded on **28 June 2020**

[38] Ved D, Saha D, Ravikumar K, Haridasan K. *Decalepis hamiltonii*. The IUCN Red List of Threatened Species 2015: e.T50126587A50131330. http://dx.doi.org/10.2305/IUCN.UK.2015-2.RLTS.T50126587A50131330.en. Downloaded on **28 June 2020**

[39] World Conservation Monitoring Centre. 1998. Drypetes porteri (e.T38763A10143318); Hildegardia populifolia (e.T33656A9801072); Koilodepas calycinum (e.T32008A9673093). The IUCN Red List of Threatened Species. http://dx.doi.org/10.2305/IUCN.UK.1998.RLTS.T38763A10143318.en. Downloaded on 28 **June 2020**.

[40] Rehel S. 2011. *Lindernia minima*. The IUCN Red List of Threatened Species 2011: e.T177197A7387587.http://dx.doi.org/10.2305/IUCN.UK.2011-1.RLTS.T177197A7387587.en. Downloaded on **28 June 2020**.

[41] CAMP Workshops on Medicinal Plants, India. *Pterocarpus santalinus*. The IUCN Red List of Threatened Species 1998: e.T32104A9679328. http://dx.doi.

org/10.2305/IUCN.UK.1998.RLTS.T32104A9679328.en. Downloaded on 28 July 2020.

[42] Balachandran N, Rajendiran K. Disjunct distribution of five endemic plants from the tropical dry evergreen forest of Tamil Nadu, India. Ind. J. Plant Sci. 2014;**4**(3):15-21

[43] Thothathri K. Fascicles of Flora of India 8 Derris. Botanical Survey of India, Calcutta; 1982

[44] Prabhu Kumar KM, Binu T, Sreeraj V, Indira B, Rajendran A. Critical notes on the occurrence of *Dipcadi montanum* (Dalz.) baker (Hyacinthaceae) in South India. Science Research Reporter. 2013;**3** (2):120-123

[45] Henry AN, Kumari GR, Chitra V. Flora of Tamil Nadu, India. Series 1: Analysis, Vol 2. Botanical Survey of India, Coimbatore. 1987

[46] Walter KS, Gillett HJ, editors. (1997). IUCN Red List of Threatened Plants. Compi. World Conservation Monitoring Centre. Gland: IUCN - The World Conservation Union; 1998

[47] Rao BRP, Babu MVS, Reddy AM, Sunitha S, Narayanaswamy A, Lakshminarayana G, et al. Conservation status of *Hildegardia populifolia* (Roxb.) Schott & Endl. (Malvaceae: Stervulioideae: Sterculieae), an endemic of southern peninsular India. Journal of Threatened Taxa. 2011;**3**(8): 2018-2022

[48] Rao DM, Pullaiah T. Ethnobotanical studies on some rare and endemic floristic elements of Eastern Ghats-Hill ranges of south east Asis. India. Ethnobotanical Leaflets. 2007;**11**:52-70

[49] Sarcar MK, Status AB. Botanical description, Natural distribution zone, Propagation practices and Conservation efforts of Hildegardia populifolia (Roxb.) Schott & Endl.- A threatened tree species of Dry Tropical Forests in India. Indian Forester. 2002:757-770

[50] Madhavachetty K, Sivaji TRK. Flowering plants of Chittoor District, Andhra Pradesh, India. Tirupati Students Offset Printers; 2018

[51] Jaikrishnan. Phytosociology and germination study of endemic and endangered species: Sterculia populifolia. MSc. Dissertaion. Botany Department. KMCPGS; Puducherry. 2018

[52] Nayar MP. *"Hot Spots" of Endemic Plants of India*. Palode, Thiruvananthapuram: *Nepal and Bhutan.* Tropical Botanic Garden and Research Institute; 1996

[53] Balachandran N. Perspectives of Plant Diversity in Tropical Dry Evergreen Forest along the Coromandel Coast of Tamil Nadu and Pondicherry. PhD Thesis: Pondicherry University; 2016

[54] Balasubramanian K. Biotaxonomical studies of Marakkanam RF, Coromandel coast. PhD thesis. Annamalai University. 1987

[55] Parthasarathy N, Selwyn MA, Udayakumar M. Tropical dry evergreen forests of peninsular India: Ecology and conservation significance. Trop. Conserv. Science. 2008;**1**(2):89-110

[56] Praveen Kumar CK. Plant biodiversity and biocultural perspectives of ten sacred groves in Cuddalore districts on Tamil Nadu. In: South India. Ph.D Thesis: Pondicherry University; 2011

[57] Ramanujam MP, Ganesan T, Kadamban D, Kumaravelu G, Devaraj P. Flora of Sacred Groves of Puducherry (a Pictorial Guide), Forest Department, Puducherry; 2007: 186 pp

[58] Ramanujam MP, Kadamban D. Rare, endangered and threatened plants

occurring in the sacred groves of Pondicherry bioregion. In: National symposium on Emerging Trends in Plant Sciences. Dept. of Botany, Sri Venkateswara University, Tirupati, 1999; Abstract 5, p 43.

[59] Isik K. Rare and endemic species: Why are they prone to extinction? Turkish Journal of Botany. 2011;**35**: 411-417

[60] Nair NC. The Southern Western Ghats; a Biodiversity Plan Studied in Ecology and Sustainable Development. Vol. 4. Indian National Trust for Art & Cultural Heritage INTACH, New Delhi; 1991

[61] Gopalan R, Henry AN. Endemic plants of India, camp for the strict endemics of Agasthiyamalai hills, southern Western Ghats. Bishen Singh Mahendra Pal Singh: Dehra Dun; 2000. 476 pp

[62] Nayar MP, Ahmedullah M, Raju DCS. Endemic and rare plants of eastern Ghats. Indian J. Forestry. 1984;**7**(1): 35-42

[63] UNEP. Global Biodiversity Assessment. Cambridge: Cambridge University Press; 1995

[64] Pitman NCA, Jorgensen PM. Estimating the size of the worlds threatened Flora. Science. 2002;**298**:989

[65] Areendran G, Rao P. Vegetation Types of the Southern Eastern Ghats – A Remote Sensing Perspective. New Delhi, India: WWF India; 2006

[66] Jain SK, Rao RR. An Assessment of the Threatened Plants of India. Botanical Survey of India, Kolkata. 1983

Section 2

Conservation

Chapter 6

Community Conservation

Amritpal Singh Kaleka and Gagan Preet Kour Bali

Abstract

Floral and faunal diversity represents the health of an ecosystem. Increase in the number of endangered plants acts as an alarming sign of ecosystem's imbalance. The ecological failure pose threat to our own health, thus by saving endangered species our own health is being saved. Government, non-profit international organizations, local communities and individuals are working together to protect and restore population levels. Biological Diversity Act (2002) for conservation of biodiversity is a landmark effort by Indian government as it provides mechanisms for knowledge, sustainable use of components of biological diversity and fair and equitable sharing of the benefits arising out of the use of biological resources. The various awareness campaigns have been conducted for local communities with regard to the conservation of endangered species. Both in-situ (on site) and ex-situ (off site) conservation strategies target critical habitats under continuous threat of extinction. Conservation programmes that centred mainly on the local masses which completely depend upon the environment including forests, lakes and wildlife for their needs truly showcase the leadership of local and indigenous communities in protecting biodiversity. The rights of local communities in decision making must be recognized and supported through clear laws and regulations. Sacred groves, a legacy of prehistoric traditions of nature conservation act as an ideal centre for biodiversity conservation. Besides providing vital ecosystem services to people, these are of immense ecological significance. Community conservation is the need of the hour in terms of conserving biodiversity at ground level.

Keywords: awareness, communities, conservation, ecosystem, regulations

1. Introduction

The plant and animal species form the foundation of a healthy ecosystem. Endangered species give the indication that the ecosystem is slowly falling apart/ degraded. A loss of single species triggers the loss of other species within the ecosystem. The fast rate of extinction and species introduction pose a major threat to our biodiversity. Over 50% of species face extinction during five mass extinctions that occurred in the past 500 million years [1, 2] and now at present we are in the opening phase of sixth mass extinction [3] which is predicted to be human impacted [4]. Plants are significantly important for conservation of biodiversity both from ecological and biological point of view. Unsustainable harvesting and habitat degradation are posing tremendous threat to plant diversity. Cure for diseases eventually comes from plants. Over 50% of the 150 most prescribed medicines were originally derived from plants or other natural products. According to the world conservation and monitoring Centre (WCMC) more than 8000 tree species are estimated to be endangered worldwide (www.unep-wcmc.org.), however, between 22% and 47% of the world's plants are predicted to be endangered [5]. So, all the plant species must be protected before they are lost forever from nature's medicine cabinet.

Plants which are scare as the total population of the species have few individuals or are restricted to a narrow geographic range are known as rare plants [6]. A non-profitable conservation organization "Nature Serve" provides the scientific information and tools needed to help and guide effective conservation action and has developed a consisted method for evaluating relative imperilment of species emphasizing rare species. A variety of factors have been summarized and assessed based on the best available information i.e.,

- Total number and condition of occurrences (e.g., populations)
- Population size
- Range extent and area of occupancy
- Short and long term trends in the above factors
- Scope, severity and immediacy of threats

"Nature Serve" has provided ranks (**Table 1**) for species based upon the status of their availability and the extent of threat possessed by the plants.

To halt the rapid loss of biodiversity (plants & animals), the Endangered Species Act (ESA) was passed in 1973. It is referred as "Crown jewel" of United Nation's environmental laws and one of the world's strongest species protection laws [7]. The ESA is uniquely effective piece of environmental legislation as it does not demand a cost benefit analysis before action.

In India, the Biological Diversity Act (2002) for conservation of biodiversity strongly recommends the creation of Biodiversity Management Committees (BMC) at village level. It also provides mechanism for declaration of the areas being conserved for agricultural or wildlife biodiversity as Biodiversity Heritage Sites (BHS). Conservation communities represent the sustainable model of community development, a new approach providing alternative to conventional development approaches. It creates sustainable employment opportunities for the local inhabitants of the community and the surrounding region [8].

Rank	Definition
G_1	Critically Imperiled- At very high risk of extinction due to extreme rarity (often 5 or few population); very steep decline
G_2	Imperiled- At high risk of extinction due to very restricted range; very few populations (often 20 or fewer); steep decline
G_3	Vulnerable- At moderate risk of extinction due to restricted range; relatively few populations (often 80 or fewer), recent and widespread declines
G_4	Apparently secure- Uncommon but not rare; long term decline
G_5	Secure- Common; widespread and abundant

Table 1.
Nature Serve Global Conservation Status Ranks for Species (G-ranks) [6].

2. Initiatives of government authorities

During last three decades, the Department of Biotechnology (DBT), GOI has undertaken a programme for conservation of 156 threatened plant species which has been successfully compiled and reviewed. A successfully tested protocol following an integrated approach for threatened species conservation is recommended for future conservation action.

Traditionally, community conserved areas throughout the country such as sacred groves and other community based protected areas have been providing in-situ conservation opportunities for the endangered and threatened species. Government has taken initiative for prioritization of threatened species and conservation actions largely on flagship faunal species. The successful in-situ conservation measures for plant groups are mainly Orchid Sanctuary at Sessa in Arunachal Pradesh and Rhododendron Sanctuary at Singba in Sikkim. The only means of ex-situ conservation of threatened plant species include field germplasm banks and institutional botanic gardens. Such programmes generally focus on ex-situ conservation of medicinally and economically important plants, and on-farm conservation of agricultural crops. All these activities are largely mentored and executed by the Ministry of Environment, Forests and Climate Change and Ministry of Agriculture, GoI. Under Umbrella Species Conservation programmes targeting big mammals, threatened plants get conserved. A protocol has been formulated for conservation of threatened plant species as under [9].

a. Through ecological niche modeling (ENM), population inventory, characterization and mapping.

b. To determine the conservation status, minimum viable population size and to assess extinction risk, meta-population modeling of selected species populations.

c. Identification of factors responsible for depleting species populations and developing a species specific recovery strategy to remove detrimental factors.

d. To identify species with greater diversity for genetic enrichment based on source-sink concept, the selected species populations undergo molecular characterization.

e. Characterization of active principles in selected species in different habitats/populations.

f. For mass multiplication, to standardize the macro and micro-propagation techniques.

g. To address the regeneration failure, the reproductive biology of the selected species is studied in detail.

h. Production of planting materials for reintroduction of the species in the areas identified through ENM.

i. Establishment of field gene banks and herbariums at appropriate ecological zones.

j. MoU with Forest Department and communities, and reintroduction with post-introduction monitoring protocol.

3. Why community-centred conservation

The only strategy to protect biodiversity is the conservation enclosures which restrict the access of communities to nature. Indigenous controlled territories are crucial to sustain habitats rich in biodiversity and the cultural knowledge that comes with those environments.

To achieve global targets of protecting biodiversity, the leadership of local and indigenous communities plays a significant role. Community conservation centers around the people that know and depend on their environment including forests, lakes and wildlife [10]. The rights of local communities to make decisions about their own resources must be recognized and supported through clear laws and regulations. The five principles have been framed by Global Biodiversity Framework 2020 through which governments and conservation partners can support community-centered conservation [11].

1. **Build trust based networks of people to collaborate for conservation:** Working together of governments, communities, environmental NGOs or other partners has positive influence on conservation outcomes due to increase in levels of trust.

2. **Promote equity and Gender equality:** The equity and equality for different groups like vulnerable communities and women are protected by International human rights. Women should be fully supported for developing strategies for the solution to the conservation dilemmas they face.

3. **Support reconciliation and redress:** Strict conservation enclosures over social injustice in conservation like land grabs etc. are characterized by forced displacement and exclusion. A conservation agenda across all levels i.e., local, regional, national and international should frame strategies to promote reconciliation and redress conservation injustices. The rights related to recognition of local and indigenous land and resource must be prioritized.

4. **Adopt a 'rights-based' approach:** For the success of conservation efforts in 2020 Global Biodiversity Framework, social justice is important. It is always neglected. It provides greater transparency about the access to critical resources, habitats and even the abilities of a person for participation in decision making. For local and indigenous communities, the transitions to rights-based are important as it enables local guardianship, thereby, improving conservation strategies. It requires increased collaboration from government and private sectors which still is an issue.

5. **Respect and revitalize local rules for decision making:** Local rules or customary systems are provided due respect. It also requires revitalization and support from partners. Efforts have been made to revitalize indigenous languages. The focus on customary rules led to the alignment with rights-based conservation strategies and focused on reconciliation as a pathway to conservation.

All these principles reflect the efforts of various international organizations along with that of government initiatives that supported community-centred conservation. These principles reflect the realities of communities (local and indigenous) as a whole and their aspirations (**Figure 1**).

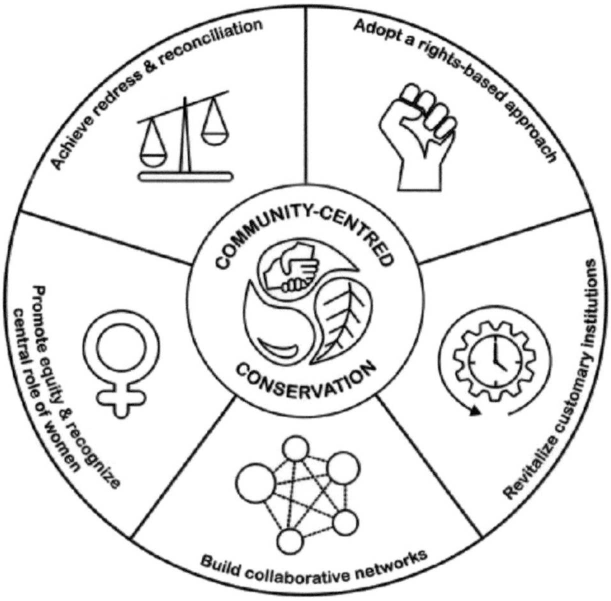

Figure 1.
Governance principles and community-centred Conservation.

4. Conservation strategies for endangered plants

The endangered plants are considered as jewels of our natural heritage and are best conserved, managed and protected by experts in collaboration with local communities. In India, the endangered species of plants (**Table 2**) are in urgent need of conservation as these are under continuous threat of extinction. The geographical areas with high number of endangered plants and large potential to protect those species should be given due attention and then the resources of endangered plants should be conserved [15].

The endangered plants which thrive in their natural habitats i.e., forests and grasslands are best conserved by maintaining healthy native habitat. The periodic monitoring is the only action taken to ensure that the plant populations are still thriving. It provides long term existence as the decreasing populations of plants can be survived at initial stages. Periodic monitoring plays a significant role in detecting early signs of decline and degradation in endangered plant populations and their habitat. Habitat restoration is possible by simple practices like removal of weeds and for those plants which rely on disturbance-dependent ecosystems like fire, wind or water.

Some plants are endemic and restricted to special soil types and other topological factors which are impossible to recreate. So, the conservation of viable habitat and maintenance of natural ecological processes are mandatory. Example: Limestone, Dolomite Mountains of San Bernardino County, California is home to plant species like Cushenbury milk-vetch (*Astragalus albens* Greene, 1885).

Pollinator decline acts as a major cause for the rarity of the plant species. In such cases, the propagation of endangered plants is carried out off-site and re-plantation of seedlings in their natural habitats is preferred. Reintroduction of pollinators in the specified habitats is also monitored, e.g., Clay phacelia (*Phacelia argilacea* Atwood, 1973).

For the conservation of endangered plant species, various in-situ and ex-situ measures have been adopted [16, 17]. Both of the practices target critical habitats, i.e., specific geographical areas which form the basis for the conservation of an endangered or threatened species. In-situ techniques focused on the prevention of total loss of the species and provide new plants for population re-establishment or habitat restoration. Botanical gardens using the system of Botanic Gardens Conservation International mainly constitute the ex-situ method for maintenance of species [18, 19].

Plant	Family	Common name	Region	Status
Acer osmastonii Gamble	Aceracea	Osmaston's Maple	West Bengal	Endangered
Actinodaphne lawsonii Gamble	Lauraceae	Not Known	Karnataka, Kerala	Endangered
Amentotaxus assamica Ferguson	Taxaceae	Assam Catkin yew	Arunachal Pradesh, Assam	Critically Endangered
Belosynapsis vivipara (Dalzell) C.E.C. Fisch.	Commelinaceae	Spiderwort	Madhya Pradesh	Endangered
Buchanania bariberi Gamble	Anacardiaceae	Not Known	Kerala	Critically Endangered
Calanthe anthropophora Ridley	Orchidaceae	Christmas orchid	Meghalaya	Endangered
Cayrata pedata var. glabra Gamble	Vitaceae	Kattuppirandai	Tamil Nadu, Kerala	Critically Endangered
Ceropegia odorata Nimmo Ex. J. Graham	Apocynaceae	Jeemi kanda	Gujarat, Melghat Tiger, Rajasthan, Salsette Island	Endangered
Cissus spectabilis (Kurz) Planch	Vitaceae	Not Known	Sikkim, West Bengal	Endangered
Commiphora wightii (Arnott) Bhandari	Burseraceae	Indian Bdellium tree, Guggul	Gujarat, M.P, Maharashtra, Rajasthan	Critically Endangered
Cycas beddomei Dyer	Cycadaceae	Beddomes' cycad	Andhra Pradesh	Endangered
Decalepis hamitonii Wight & Arn.	Apocynaceae	Swallow root	Peninsular India	Endangered
Dicliptera abuensis Blatt.	Acanthaceae	Not Known	Rajasthan	Endangered
Eugenia argentea Bedd.	Myrtaceae	Not Known	Kerala	Endangered
Ilex khasiana Purk.	Aquifoliceae	Not Known	Shillong peak in Meghalaya	Critically Endangered
Impatiens nilagirica Fischer	Balsaminaceae	Not Known	Tamil Nadu	Endangered
Kingiodendron pinnatum (DC.) Harms.	Fabaceae	Malabar Mahagony	Kerala, Karnataka, Tamil Nadu	Endangered
Ophiorrhiza incamata Fischer	Rubiaceae	Not Known	Kerala	Endangered
Pterocarpus santalinus Linn. f	Fabaceae	Red sandalwood	South Eastern India	Endangered
Renathera imschootiana Rolfe	Orchidaceae	Red vanda	Manipur, Nagaland, Mizoram	Endangered

Table 2.
Endangered plants in India [12–14].

The recent advancement in plant biotechnology permits the maintenance and propagation of endangered species. It allows high added value compounds to be obtained in a sustainable and ecological way in plant cell and organ cultures without need to harvest plants under pressure in their native habitat. Nowadays, it is easier to investigate the population structure, genetic variations and recent demographic events in threatened species using population genomic approaches. A genomic approach provides detailed account of the present and past demographic parameters, phylogenetic issues and molecular basis for inbreeding. It also helps to understand genetic diseases and provides information about the mechanisms related to low fitness with low genetic variations. Genetic and environmental methodologies are integrated into conservation biology and used to design fast monitoring strategies [20].

Seed banks can be maintained through desiccation or by cryopreservation (−196°C) [18, 21]. The Royal Botanic Gardens of Kew is host to the world's ex-situ collection of seeds from wild flowering plants. The plant species which cannot be conserved through conventional seed banking need other treatment as dormancy of rupture [19]. In some plants the seed production is very low so seed storage is impossible. Therefore, the strategies which induce physiological and mechanical seed dormancy rupture must be carried out to cause a positive response on seed germination [21].

Micro propagation is a method of producing certain plants in much larger numbers than by natural methods. It results in large number of clones of the original plant via tissue culture techniques, the clones can be further used to cryopreserve or reestablish the plant populations [21]. It is the only viable option for ex-situ conservation of those plants which cannot be conserved through conventional seed banking [18].

There are many ways to develop strategies for conservation of endangered plant species. In order to protect plant species and stop further plant declination, get involved in the conservation strategies and programmes [22].

5. Main objectives of communities involved

Communities embodied a wide range of objectives for which they ultimately conserve biodiversity. It is not necessary that Biodiversity conservation is the main objective behind the action. Some of the objectives of the communities involved are as under [23]:

a. **Resource enhancement/maintenance:** A serious scarcity of timber, fuel wood, fodder, medicinal plants etc. faced by communities initiated an effort towards conservation along with sustainable management of ecosystem. Conservation is also initiated by local inhabitants to ensure the continuous availability of resources for future generations. Examples include: Van Panchayat forests in Uttarakhand; Kailadevi Wildlife Sanctuary in Rajasthan; Jardhargaon in Uttarakhand; CFM (Community Forest Management) villages in Orissa and West Bengal etc. Resource enhancement and maintenance is one of the highest motivational factors for communities to start conservation.

b. **To counter ecological threats:** Ecological threats like frequent landslides, drought conditions, reduced soil fertility, non availability of water and other natural calamities led the communities to take action. Examples: Hunsur village in Karnataka and Konark-Balukhand in Orissa.

c. **To counteract developmental threats:** The natural resources are under continuous and alarming threat from commercial and developmental projects. Community's livelihood completely depends upon these resources. So as to safeguard them, the community conservation is initiated. Examples: Chipko movement (against timber logging) in Uttarakhand and Mendha village (against dams and paper mill) in Maharashtra.

d. **Religious sentiments and Cultural ethics:** India is a land of divine entity and home of saints. The religious sentiments of people are associated with plant/animal species, sacred landscapes and rivers. To protect these religious entities, communities bound themselves to conserve biodiversity. Examples: Sacred groves (**Table 3**) like Ajeevali village in Maharashtra, sacred landscapes of Sikkim, sacred ponds and forests of Uttarakhand etc. The sacred grooves include relic forest patches that form important repositories of forest biodiversity and provide abode to numerous plant and animal species. An inextricable link between the present society and past in terms of biodiversity, culture, religious and ethnic heritage exists in sacred grooves [25]. These grooves act as ideal centre for biodiversity conservation. Sacred grooves are found all over India especially in those regions where indigenous communities inhabit.

Our country is a place where multiple customs and traditions not necessarily linked to religious sentiments are being practiced. These customs and traditional beliefs also inculcate the sense of responsibility in common folks towards the protection of resources. Examples: Community land use in North-East India, heronries in villages like Kokkare Bellure in Karnataka etc.

S. No	States	No. of Documented Groves
1.	Andhra Pradesh	750
2.	Arunachal Pradesh	58
3.	Assam	40
4.	Chhattisgarh	600
5.	Gujarat	29
6.	Haryana	248
7.	Himachal Pradesh	5000
8.	Jharkhand	21
9.	Karnataka	1424
10.	Kerala	2000
11.	Maharashtra	1600
12.	Manipur	365
13.	Meghalaya	79
14.	Orissa	322
15.	Rajasthan	09
16.	Sikkim	56
17.	Tamil Nadu	448
18.	Uttaranchal	01
19.	West Bengal	673
Total = 13,720		

Table 3.
Sacred Groves in India [24].

6. Global scenario of community conservation

Conservation of biodiversity is a global issue, important for humanity as a whole [26]. Conservation is typically a complex system as the natural environment itself is a complex adaptive system [27]. The global as well as the local objectives of conservation of resources are fulfilled by community-conservation strategies and programmes. The global scenario of community based conservation is being supported by a case study of indigenous communities of Kenya.

Maasai and Rendille Communities of Kenya: A case study

The two indigenous communities namely Maasai and Rendille in Kenya played a significant role in conservation of biodiversity. Maasai people are basically nomadic community that co-existed with wildlife. The land resources along with the flora and fauna were under continuous developmental threat that initiated the community conservation.

The Maasai from Trans Mara, Narok country and the Rendille from Kargi, Kamboye, Korr and Logoloho of Marsabit country carried out conservation strategies after witnessing the increasing biodiversity loss. They created and managed 6,000 acres of land under Nyekweri Forest Kimintet trust so as to conserve forests and wildlife. Massai are considered as unique for their cultural prohibition against hunting of wild animals except in extreme conditions [28]. The community members asserted that their area is an important biodiversity hotspot with indigenous forests hosting sacred sites; sheltering wildlife; trees providing food, fodder, fuel and medicinal plants. During assessments, community men from Nyekweri Forest Kimintet trust identified key external threats such as the erosion of cultural values governing conservation, privatization of lands and loss of traditional, ecological knowledge and deforestation due to charcoal burning and overgrazing to the existing biodiversity [29].

Women of these tribes rarely participate in decision making bodies. They are the stake holders of traditional knowledge and values and act as inter-generational transmitters of knowledge [30]. The community members undertake reforestation initiatives and conserved their area. It helped in restoration of degraded areas with the indigenous tree species. The community members of trust proposed solutions to threats including support for the community as [29]:

- Exchange visits to other wildlife conserved areas for practical learning and adaptation of best practices.

- Strengthening linkage with country and national government environmental authorities like Kenya Forest Service (KFS) and Kenya wildlife service (KWS).

- Financial support for management and reforestation initiatives.

- Raising awareness to strengthen the role of women in conservation by creating women's networks for conservation.

Alias Morindat (International Institute of Environment & Development) has advocated the significant role of local people in conservation strategies and put forth as,

"Indeed people should not be seen as objects or empty vessels' but rather as drivers of their destinies and masters of their own development."

7. Limitations of community conservation

Communities that are actively involved in the conservation of an area that provides livelihood to them are under continuous threat from internal as well as external factors thereby restricting them to get social, administrative and legal recognition. Community conservation has certain, unavoidable limitations that require urgent thought process for better understanding and resolution. Community conservation may not necessarily address the issue of overall biodiversity conservation as species that are useless or undesirable to the community may not be given attention [10]. Both ecological as well as social conditions hinder the conservation programmes. The limitations should not be demoralizing but emphasize areas where the community and other partners (NGO) have little control and need to build partnerships. It highlights the need for government partnerships. Some of the important limitations are discussed as under:

a. **Man-Animal conflict:** Due to the protection of the areas, the population of wild animals and birds increased leading to the increase in crop damage and livestock loss. This situation leads to serious man-animal conflict for the procurement of resources. In some areas, the concerned villagers have opted for reopening of regulated hunting of animals. Even the government has taken initiative to help local people in such conserved areas e.g., the Snow Leopard Conservancy in Hemis National Park in Ladakh has initiated a programme aimed at helping local people in reducing damage to livestock caused by the snow leopard and help them in getting adequate and timely compensation for the incurred damages [31].

b. **Lack of Monitoring and Evaluation:** For the better and expected results the working of the communities should be properly and timely monitored. But there are few community conservation efforts that are monitored and evaluated for assessment of social and ecological impacts. The internal monitoring system in maximum of the involved communities is lacking. The trends and changes observed through monitoring and evaluation are purely technical and need critical assessment.

c. **Lack of baseline information:** The basic information about the flora and fauna of the area is not available. Communities rely on the oral histories and information narrated by the older people which are not authenticated. The youth of the area should be motivated and trained to document the flora and fauna of the local area so as to implement conservation strategies. Environment education is essential for the future of conservation and educational components are thus critical for sustainable community-based conservation initiatives.

d. **Limited capacity:** It is not possible for communities to manage finances, manage ecosystems and monitoring and evaluating results. They rely on others for administrative skills. The capacity building programmes for various government officials including forest officers and officials of other departments, NGO's should be organized to sensitize them to work together for the common cause of conservation. They should play a role of support, facilitation, guidance and mentorship. Community conservation is a slow process as it is mainly a social process and has to progress taking into consideration various issues. The implementing agencies along with the collaborated NGO's have to work in accordance with each other to achieve the targets at stipulated time.

The success of community conservation mainly depends upon the empowerment of the whole community. Vulnerable communities which are affected by ecological changes should be involved actively in conservation practices. Conservation reforms lead to other reforms in that area such as empowerment, gender equity, education, cooperation etc. Thus, community-centred conservation is not an isolated entity.

8. Conclusion

Eliana Elias- National geographic Explorer has accurately mentioned, "A Community in balance leads to a planet in balance". Community based conservation is being increasingly recognized as a major global force in the protection and sustainable management of ecosystems and species [32]. Plants are ecologically and biologically an important constituent of ecosystem. Forests host sacred sites, provide food, fodder, fuel, medicinal plants etc and are home to diverse flora and fauna. Forests are socially and culturally significant for those who have conserved them through customary laws. But the present scenario of continuous habitat destruction, introduction of invasive species, decline in natural pollinators and most devastating the threats from developmental projects led to the species extinction and increased number of endangered species. The over exploitation of medicinal plants aids the destruction caused to rare plants.

Despite of strict implementation of conservation laws which include prohibition of the collection of endangered plants from grasslands and forests; rigorous punishment to the defaulters etc. are still the prime motive i.e., conserving flora of a region is not fulfilled effectively. Conservation should be people centred and address the diverse livelihood of community along with the conservation of biodiversity. Sacred grooves are legacies of ancient traditions of nature conservation forming a strong linkage between present and past in terms of biodiversity. Biodiversity rich sacred grooves possess immense ecological and social significance. These grooves thrive well and automatically conserved at those places where indigenous communities inhabit. Communities hold the key to expand conservation strategies so as to safeguard biodiversity although it facilitates recovery but does not consider the requirements of each species [33]. By developing involvement with communities and engaging in development and conservation strategies, win-win solutions can be created [29]. Women although considered as inter generation transmitter of knowledge and values should be involved in decision making process i.e., providing gender equity and thereby emerging as allies in affecting change in the society. A sustainable earth cannot be created alone. Communities must be advocated for conservation and collaboration.

Author details

Amritpal Singh Kaleka* and Gagan Preet Kour Bali*
Department of Zoology and Environmental Sciences, Punjabi University,
Patiala, Punjab, India

*Address all correspondence to: apskaleka@gmail.com and gaganviren@gmail.com

References

[1] Myers N. Mass extinction and evolution. Science. 1997;**278**:597-598

[2] Erwin DH. The end and beginning: recoveries from mass extinctions. Trends in Ecology and Evolution. 1998;**13**:344-349

[3] Myers N. Questions of mass extinction. Biodiversity and Conservation. 1993;**2**:2-17

[4] Eldredge N: The sixth Extinction. 2008. Available online: http://www.actionbioscience.org/newfrontiers/eldredge2.html#primer.

[5] Graham S: Global Estimate of Endangered Plant Species Triples. 2002. Available online: https://www.scientificamerican.com/article/global-estimate-of-endang/

[6] Rare Plants. Available online: https://www.fs.fed.us/wildflowers/Rareplants/

[7] Conservation of Threatened plants of India. 2019. Available online: dbtindia.gov.in/conservation-threatened-plants-India

[8] Randall A: Resource economics: An economic approach to natural resources and environmental policy. 2nd ed. New York, USA: John Wiley and Sons. 1987; p. 448

[9] Christopher K: Inside the effort to weaken the Endangered species act. 2019. Available online: nationalgeographic.com/news/2017/05/endangered_speciesact/

[10] Armitage D, Muhl EK, Sowman M, Mbatha P, Rice W: Why communities must be at the heart of conserving wildlife, plants and ecosystems. 2020. Available online: https://theconversation.com/why-communities-must-be-at-the-heart-of-conserving-wildlife-plants-and-ecosystems-132416.

[11] Armitage D, Mbatha P, Muhl E.K, Rice W, Sowman M: Governance principles for community-centred conservation in the post-2020 global conservation framework. Conservation Science and Practice. 2020; 2: 1-18. https://doi.org/10.111/csp2.160

[12] Nayar M.P, Sastry A.P.K: Red Data Book of Indian Plants. 1990; vol. 1-3. Botanical Survey of India, Calcutta.

[13] Available online: bsienvis.nic.in/Database/RedlistedPlants-3940.aspx.

[14] Available online: nelda.org.in/blog/category/importance-of-trees/7/endangered-trees-species-in-India-which-may-soon-go-extinct.

[15] Faleiro FV, Machado RB, Loyola RD. Defining spatial conservation priorities in the face of land-use and climate change. Biological Conservation. 2013;**158**:248-257

[16] Troyer CM, Gerber LR. Assessing the impact of the U.S Endangered species Act recovery planning guidelines on managing threats for listed species. Conservation Biology. 2015;**29**(5):1423-1433

[17] Havens K, Kramer AT, Guerrant EO. Getting plant conservation right [or not]: the case of the United States. International Journal of Plant Sciences. 2014;**175**(1):3-10. DOI: 10.1086/674103

[18] Cibrian-Jaramillo A, Hird A, Oleas N, Helen M, Meerow AW, Javier FO, et al. What is the conservation value of a plant in a Botanic Garden? Using indicators to improve management of ex-situ collections. The Botanical Review. 2013;**79**:559-577

[19] Hilton-Taylor C.A: History of IUCN DATA Book and Red list. 2012; 5-11.

[20] McDonald AJS, Davies W. Keeping in touch: responses of the whole plant to deficits in water and nitrogen supply. Advances in Botanical Research. 1996;**22**:229-300

[21] Pimm L, Joppa N. How many plant species are there and at what rate are they going extinct. Annals of the Missouri Botanical Garden. 2015;**100**:170-176

[22] Goleniowski ME, Cusido R. Endangered plant species Recovery. JSM Biology. 2017;**2**(2):1013-1019

[23] Pathak N. Community conserved Areas in India: An overview. Kalpavriksh. 2009:45-94

[24] Akshay KR, Sudharani N, Anjali KB, Deepak TM. Biodiversity and strategies for conservation of rare, endangered and threatened medicinal plants. Research and Reviews: Journal of Pharmacognosy and Phytochemistry. 2014;**2**(3):12-20

[25] Khan ML, Khumbongmayum AD, Tripathi RS. The sacred grooves and their significance in conserving biodiversity An overview. International Journal of Ecology and Environmental Sciences. 2003;**34**(3):277-291

[26] Capistrano D, Samper C, Lee M.J, Raudsepp H.C: Ecosystems and Human Well Being. 4th Ed. Millennium Ecosystem Assessment and Island Press, Washington DC. 2005; p 53.

[27] Levin S. A: Ecosystems and the biosphere as a complex adaptive systems. Biodiversity. 1998;**1**:431-436

[28] Llamazers AF, Western D, Galvin KA, McElwee P, Cabeza M. Historical shifts in local attitudes towards wildlife by Massai pastoralists of the Amboseli Ecosystem (Kenya): Insight from three conservation psychology theories. Journal of Nature Conservation. 2020;**53**(125763):1-11

[29] Kaptoyo E: Report of the community conservation Resilience Initiative in Kenya. Global Forest Coalition. 2018; 1-24. Available online: http:// globalforestcoalition.org/resources supporting-community-conservation/

[30] Lichtenfeld L.L, Naro E.M, Snowden E: Community conservation, and collaboration: A framework for success. National Geographic society, Washington D.C., United States and African people & wildlife, Arusha Tanzania. 2019; 1-48.

[31] Jackson R, Wangchuk R. A community based approach to mitigating livestock depredation by snow leopards. Human Dimensions of Wildlife. 2004;**9**:307-315

[32] Kothari A, Camill P, Brown J: Conservation as if people also mattered: Policy and Practice of community based conservation. Conservation & Society. 2013; 11: 1-15. Available online: https:// www.conservation and society.org/text. asp? 2013/11/1/1/110937.

[33] Ravikanth G, Jagadish MR, Vasudeva R, Shaanker RU, Aravind NA. Recovery of critically endangered plant species in India: need for a comprehensive approach. Current Science. 2018;**114**(3):504-510

Chapter 7

Germplasm Conservation

Sameer Quazi, Tanya Golani and Arnaud Martino Capuzzo

Abstract

With the increase in risk of extinction of various plants, the trend has been shifted to employment of many biotechnological techniques for preservation of genetic resources of plant and is the area of research which needs to be revolutionized after a specific time period because it allows the production and selection of crop varieties with desirable characteristics during breeding process such as improved fuel, food and health facilities. Having an immense research in conservation of non-threatened species, there is a small collection of knowledge available for conservation of endangered ones. This chapter aims to highlight the various techniques in germplasm conservation of endangered or the species which are at extent of extinction and also the future directions in this field. In developing countries where most of agriculture depends upon food crops, the maintenance of genetic variation is of immense importance. On farm conservation provides the best example of preservation and evolution based on genetic variability which can occur ex-situ and in- situ environment in farms or gene bank. So, it presents the best option for conservation or maintenance of ecosystem and biodiversity which ensures survival of endangered species via germplasm. The most point to consider is that germplasm or genes have to be conserved instead of genotype. In situ conservation involves preservation of plant crops in the field condition in ecosystem where plant is adopted to grow in order to maintain self –sustaining process in natural ecosystem. Similarly ex-situ involve the collections of seed banks of genes collected from plant under natural conditions to produce desirable varieties or from tissue culture in laboratory also referred as in-vitro methodology. In –vitro techniques include cryopreservation which include freezing at much lower temperature than that of freezing point i.e. -196 °C in liquid nitrogen for preserving species which are near to extent of endangerment. Cold storage and storing at lower temperature provides best opportunity for protection against damage caused by rapid freezing. Germplasm exchange has become now a usual practice ensuring exchange of varieties between cultivated and wild types as for example in potatoes specie etc. DNA as well as gene or seed banks provide molecular sources for conservation at biotechnological level. The techniques of introgression and incorporation are basic approaches for germplasm conservation. So there is need to revolutionize and practice germplasm conservation for fulfilling future needs being aimed at conserving endangered or threatened species from conservation hotspots.

Keywords: germplasm, threatened species, gene resources, cryopreservation, introgression, incorporation, endangered species, tissue culture technique, seed bank, gene bank, gene pool, and breeding technology

1. Introduction

Conservation of plant genetics is one of the main areas which is to be refined and revolutionized again and again with knowledge. The phenomenon of conservation is helpful in the maintenance of the genetic basis being needed for breeding. This allows the production and selection of varieties with desirable characteristics in crops, which later can be used for purpose of feeding, fuel, and health sectors [1]. Germplasm is the plant's genetic resources such as tissues or cells which are being preserved for purpose of obtaining desired breeding characteristics. These resources are obtained from gene banks, plants grown in nurseries, and laboratory culture. The collection of germplasm usually ranges from wild species to genes which are supposed to capture traits of plants as a result of natural selection [2].

A germ is defined as the collection of genetic resources for an organism. In the case of plants, the germplasm is stored or preserved in form of seeds or trees in the nursery. So, it is the living tissue from which new varieties of plants can be grown i.e., it can be the seed from which the whole plant can be grown because it contains all the genetic makeup or information required for resources of the diversity of plants. Plant germplasm is a spice of generic materials needed by breeders to develop new varieties. This includes seeds, leaves, stems, pollen, and cultured cells. So it provides the necessary raw material to develop the commercially valuable varieties of plants [3].

It is of prime importance in the maintenance of diversity in the biological system and the security of food. Conservation of plant resources is of great importance because most of the plant species are getting endangered with time. Genetic resources are a potential sustainable source of agricultural products i.e. efficient production of crops used as food, for the reduction of poverty and maintenance of economic conditions of the population [4]. For example, in countries like Nigeria, the major source of food is from crops sowed by simple farmers which maintain them by their efforts utilizing their resources. This involves the conservation of gene resources which preserve them for storage and usage systemically at both national and international levels [5]. Hence maintaining these species for purpose of variation in genetics is therefore of immense importance especially in the case of poor farmers who are participating in the agriculture of the country at much lower input conditions in marginal land [6].

On-farm conservation provides the best example of preservation as it is helpful in the maintenance of evolution responsible for genetic variability. Variations in genes are observed in both ex-situ i.e. in a natural environment and ex-situ in the form of gene banks obtained from laboratory culture. A huge collection of the most value able crop fields are being reserved in the gene bank and are placed in modern aseptic conditions in gene bank facilities. The variety includes collection from national and international worldwide programs i.e., NAGRAB, OSTROM, and IITTA. It also includes varieties obtained from plant genetic programmers in collaboration with a national action plan [7]. It has been revealed from a scientific investigation that about 3 lac plant species of higher plants exist in the world but only 1% of them are being utilized in the world today. About 80% of food is provided by only 8–10 crops ranging from wheat, rice species to millet, and rye. Most advancements in the field of agriculture in the present day world is based on a wide range of genetic resources possessing two types of values [8]. The plant genes and various genotypes are considered for many characteristics such as insect and pest resistance, bearing the conditions of drought, plant structure, function, and color acting as an immediate source of plant genotype conservation for desired properties. Secondly, diversity in genes or genetics ensures future requirements. Hence in turn contribute to the farming system at both local or small level and national levels [9].

Moreover, variations in genetics have also resulted in losing information in an already present generation which makes the preservation of these genes much important. Because if genes for variations are not preserved, it would lead to endangerment of plant species. International board [7] as a bank for further next generations [10]. The conservation involves

- Preservation of breeding lines
- Conservation of commercially important species
- Stock for genetics
- For direct or indirect usage of wild species which include either in form of crops or stocks of roots.

2. Need for germplasm conservation of endangered plants

There are several reasons why breeders use this technique which is as follows;

- Loss of genetic diversity among plant species.
- Humans and animals are dependent on plants for their food which means they require plants for basic food crops such as wheat, maize, etc.
- Humans also utilize plants for their social activities such as buildings construction, obtaining waxes, and perfumes, resume fibers, and therapeutics.
- Deforestation has led to the endangerment of many valuable varieties of plants which present an utmost need to preserve them [11].
- To keep the stability of the ecosystem, genetic diversity provides food prints that can be maintained via germplasm.
- It provides an esthetic value to the natural ecosystem and bio-diversity of plants [12].

3. Classification of germplasm

Based on its need, germplasm has undergone an evolution over a certain period of years in response to particular requirement including

1. Base collections
2. Backup collections
3. Active collections
4. Breeder collections
5. Working collections

To a certain extent, these collections are artificial to a much extent because some of the classes or classifications are useful for more than one reason. Hence an active number of collections were previously breeder such as for formal breeding purposes. So, the following discussion is required to explain the classes of the collection which must serve.

3.1 Base collection

It presents the method of long-term preservation of genetic variability by storing in presence of optimum conditions. For base collection, the materials are not used for distribution except with the need for replacement of material that has been lost either from active or backup collection types. It includes the most explanatory sampling method being employed for checking out variability within the species group. They are most stable in the sense that they can store the variation which arises in the natural condition. But they are also dynamic in the sense that they have some novel collected materials, some collections being produced via plant breeding and population involving genetic materials are added as they are available. In this way, the storing for many decades can be possible so the loss of variability occurs during the processes of regeneration and storing present within the acceptable limits. It is the collection under a low level of humidity at the temperature of subfreezing which must be below −150 °C to 190 °C. But some difficulties are present, which include that they cannot bear the chilling or drying temperature. So, an alternative and long-term methodology are required which includes cryopreservation within the in-vitro cultures. A huge collection at a global level is initiated with a proper guidance and help of Food and agriculture organization by designing specific agencies which serve as the base as well as back up collections for principal species in case of principle crop plants. But most agencies also vary in their ability to fulfill all the responsibilities regarding its designation [13].

3.2 Back-up collection

It supplements base collection at another location or another level. For example, laboratory at US national seed stores holds the collection of some of the duplicate backup samples of maize for the improvement of these crops. Similarly, the international research Centre of rice is abbreviated as IRR present in the Philippines for the collection of rice. So it holds collection as well as insurance for loss in primary CIMMYT and IRRI collections of crops [14].

3.3 Active collection

Active as well as base collection mostly includes the same type of materials. So, it provides the seeds and other raw materials for purpose of distribution as well as for other uses. So, it has been found that a certain collection of material is conserved for maintenance of sufficient collections of plants of each type in active collection particularly when it is required in a huge collection or amount. All the materials in this type of collection are maintained under a shorter half-life and in the response to more standards of variability. So, grow outs or techniques for replacing seed supplies in form of the active collection as compared to that available in the form of the base collection. Hence replacing the active collet ions being necessary at regular intervals is being necessary in the case of the base collection so puts the genetic association at the risk [15].

3.4 Breeders and working collection

Breeders as well as the working collection include materials being used in breeding programs and are used for the short term in nature. Breeders get knowledge from their experience such as superior performance in their local region has resulted in the favorable combinations of different alleles at an almost different genetic locus. Hence attempts for the introduction of alleles from exotic resources into adapted materials are determined to the performance in the short-term way. So, the breeder's collection includes the advanced cell lines developed in their programs in addition to professional cultivars, advanced breeding lines, and finally genetically enhanced as obtained from a breeding population in the presence of a similar type of ecological variable conditions. It is suggested from breeders that dependence on the already available stock has resulted in slow advancement towards the new technology. But modern breeders turn into exotic materials for utilization of variability in the active collections. But they obtain the variability in exotic alleles or genes from the genetically enhanced population or breeding stocks in both of them the most useful and desirable alleles have been introduced. So breeder's collection has turned ultimately to increase in proportion for genetically enhanced stock which can possess the useful alleles in the genetic backgrounds too [16].

4. Classification of the gene pool

Gene pool includes almost all the cultivars which can be obsolete or current, or wild species and their relatives which in turn contain genes available for utilization in true-breeding lines. Based on their relationship, the gene pool can be classified into three major classes;

- Primary gene pool
- Secondary gene pool
- Tertiary gene pool

4.1 Primary gene pool

It is abbreviated as GP1 and it is the form of gene pool where the crossing of two species is much easier which ultimately leads to the production of sexually fertile organisms. It includes plants or other species which upon mating produce a very closely related species which is fertile via its reproductive means. In the gene pools, the genes can be exchanged in between the two reproductive lines via arranging simple crosses or hybridization patterns. So, it is also known as Gene pool one and is of prime importance in breeding lines.

4.2 Secondary gene pools

This type of genetic material can lead to a partial type of fertility upon crossing with GP1 being referred to as the secondary gene pool. It can ultimately cross with the primary generic pool but the hybrids obtained after the process of hybridization usually produce offspring which are fertile to some extent which means that some of them are fertile while others are sterile. Transferring such genes to the primary form of the gene pool is a much difficult and laborious task and such type of genetic pool is also known as Gene pool two (GP2) [17].

4.3 Tertiary gene pool

It includes the type of genetic material which produces the ultimate sterile type of hybrids while crossing with the primary types and hence the name tertiary is given to them being abbreviated as GP3. It owns the material that can be easily crossed with the primary type of gene pool but the offspring after hybridization will produce a sterile organism. So transferring such materials to the primary gene pool is only possible in the presence of specific biotechnological techniques.

5. Activities of germplasm

There are six types of activities being related to gene resource which include;

- Collection or exploration of germplasm
- Conservation of germplasm
- Evaluation
- Documentation
- Distribution
- Utilization of germplasm

5.1 Collection of germplasm

Exploration refers to the collection of germplasm or in other words collecting the variable genetic resources from different sources and placing them at one place which is a highly scientific procedure. Collection can be done from five sources i.e., from diversity centers, gene Banks or sanctuaries, companies for seed collection, and finally through fields. Secondly, germplasm collection is done based on endangerment i.e., the species or crops which are more at the extent of extinction are preferred more as compared to others. The method of collection is done in presence of agricultural universities in collaboration with the National Bureau of genetic resources of the plant in New Delhi. For collection at the global level, it is done at the global level by International plant genetic resources being abbreviated as IPGRI with Rome and Italy [18]. The collection is done based on migration to areas of more genetic diversity, by visiting the gene bank by yourself, and finally via the exchange of genetic material. Similarly, there exist two methods for the exchange of germplasm which include random sampling involving the collection of genetic traits for both the biotic or abiotic stresses while abiotic involves collecting the different phenotypically traits. Hence both the random as well as non-random sampling methodologies are employed for collecting germplasm. Sampling size should be such that it can collect about 96% of diversity occurring in genetic traits [19]. Hence it involves the collection of 55% crop plant species of seeds per plant. Also, a wider range of habitats is sampled for obtaining maximum diversity accordingly. But there are certain drawbacks of exploration or collection such as reduction of genetic diversity due to occurrence of genetic erosion, collection from other countries or sites leads to disease condition leading to spread of weeds or pests. Moreover, it is a tedious job that requires drilling, lodging, and transport. Lastly collection from huge resources promotes problems in the collection as well as transportation. While some of the merits include the discovery of new species while exploration and also help in the preservation of certain genotypes that have become either extinct or at the extent of extinction [15].

5.2 Conservation

Conservation involves the protection of the genetic diversity of plant crops from genetic erosion which can be either ex-situ or in situ. In situ refers to conservation under natural habitat requiring establishing resources of biosphere or ecosystem for the preservation of endangered crops or plants for future usage. Following this method both wilds, as well as natural biospheres, are conserved presenting the disadvantage of covering a very small area of genotype in the case of single species, it is a much expensive methodology and also requires a proper management system. An ex-situ conservation germplasm is conserved in form of a gene bank which is a most practical application being employed under laboratory conditions. This methodology enables the preservation of whole genetic diversity in one place. Moreover, the method is in-expensive and easy to operate [20].

5.3 Evaluation

Evaluation involves the investigation or examination of genetic resources of plants based on their phenotypically, genetics, economic, biological, and chemical characteristics. It is essential for the identification of resources for the resistance, production, yield, and other quantitative characteristics. It provides all the necessary information regarding the classification of germplasm and their characterization of each of the individual germplasm attributes. It involves the requirement of a team of specialists from physiology, biotechnology, biochemistry, and entomology. For all the characters evolution is done separately and experts from IPGR, Italy. Evolution is either done infield, in the laboratory, or greenhouse. Observation is done on basis of morphological characteristics and is recorded via specific instrumentation. The characters of resistance and biotic or abiotic stresses are screened in the greenhouse. While the evaluation of biochemical characters is done on basis of conditions under laboratory. Both visual as well as instrumentation is done accordingly [21].

5.4 Documentation

Documentation involves storage, analysis, and dimension. In-plant genetic resources includes the collection, evolution, storage, and conservation of information. But now it is termed as an information system. A large collection of information is available for major crops such as maize, sorghum, wheat, and rice, etc. Till now about 7.6 million germplasm are available for the conservation of about 300 or more species. Handling of the huge collection is done via the involvement of electronic computers. For uniformity of characteristics, it involves standard characters and further descriptors for comparison in IPRGI. The information is stored in the memory of a computer and must be available at the time of need when required.

5.5 Distribution

Distributions are the most important activity for genetic resource centers. During this process, specific germplasm is supplied to the users for improvement of genetic traits and is responsible for the maintenance of conserved germplasm and its supply at a time for utilization. Distribution is the responsibility of the gene bank center's where they are maintained and being stored and to those who are engaged in specific research activities of a particular crop. The amount transferred as a sample is very small and depends upon the type of material available in raw form and also several other factors. A proper recording system is maintained and checked after the report by the user which tells the most important characteristics of association to the distributor. Germplasm is usually distributed after collection for at least two crop seasons because it is helpful in the adoption and purification of plant material [22].

5.6 Utilization

Utilization involves the employment of conserved germplasm in research and improvement programs and can be utilized in various paths mainly in three forms such as;

- As a new variety of crop
- As a sample in the hybridization of plants
- As a genetic variant allele in crop improvement.

Some of the crop varieties are made available instantaneously after their testing because in this case performance of these exotic gene lines are found to be better than that of local varieties so it will be available for usage at the commercial level. In another case, new varieties are developed based on selection done from the already present collection. In either case, some of the germplasm is not usable at all but possesses certain characteristics such as resistance, economic or wider adaptability. Transfer of such germplasm is easy because it can show cross-compatibility. The similarly wild form of germplasm is used for providing resistance to biotic or abiotic stresses and other characteristics such as strength in cotton. But it will present some further problems which include; the inability of the hybrid to survive for a large period. Sometimes the hybrid plant is unable to produce its offspring's and desirable characters get linked to undesirable ones. Hence the utilization of germplasm is a difficult task and requires special attention [23].

6. Involvement of organizations in germplasm conservation

Two of the organizations on both national as well as international level are available having an association with preservation or conservation of germplasm of plants. Thus, providing the facility of their abrupt usage when necessary by them. These include the international plant genetic resources institute located in Rome, Italy operating at the global level. Various types of institutes work and deal with the germplasm of concerned and most important crop plants. However, in India National Buraeu of Plant Genetic Resources abbreviated as NBPGR deals with a huge collection of both horticulture and agriculture crops. In addition to them, Forest research institutes deal with species living in forests and lastly Botanical survey of India located in Kolkata deals with the remaining plant species [24].

6.1 Role of IPGRI

IPGRI old IBPGR is an international scientific organization whose work in addition to other institutions is analyzed by CGIAR which is a consultative group on international agricultural research. Its main role is to conduct and organize research and also to promote and ensure collection, documentation, and utilization of these plant germplasm and it will be helpful in the collection and exchange of plant materials. It also possesses an advisory committee which helps in the collection, evaluation, and utilization of germplasm of crop plants. So it promotes global collection and conservation of all the genetic resources of plant species. It was changed from IBPGR to IPGRI in 1993 while its predecessor was established in 1974.

6.2 Role of NBPGR

This institute was established by the Indian Council of Agriculture research Centre (ICAR) in New Delhi in 1976. In India, the introduction of the plant was done in 1946 in the Division of Botany and a separate introduction was done by Dr.H.B. Singh who made a well-known achievement in the fields of the introduction of plants in India. He also arranged a large collection of germplasm of various species of plants and systemized the research in this field. In 1976, this decision of introduction of plants was revolutionized to an independent agency named NBPGR. The basic function is that it is helpful in the import and export of genetic resources of plants hence facilitating the exchange of germplasm. Also, it promotes activities of germplasm like collection, conservation, documentation, and utilization. It also organizes short term courses of collection, conservation, evaluation, and utilization of genetic resources of crops. Besides, it also guides the development of storage of plants at cold temperatures and short-term conservation of germplasm. It's also a decision about the setting of the gene for endangered species of plants [25].

7. Genetic erosion

It is also known as genetic depletion in which a limited number of genes of species that are endangered get more reduced where reproductive individuals die before reproduction with others in their low population. In a more detailed way, it is described as a loss of some alleles or genes while referring to further loss of the whole phenotypic trait or genotype. It occurs because each individual has a unique set of genes that get lost when they die before they breed. A low level of genetic diversity leads to further reduction of the genetic pool thus breeds a combination and also weakens the immune system taking the species to the level of eventual extinction. Genetic erosion is greatly observed in endangered species. Most crop species get benefits from most of the human-associated programs to keep the production viable. So in this way avoids extinction for a large time frame [26]. A small collection of populations are more vulnerable to erosion than that of a larger one. The level of erosion gets worse and is being accelerated with time-based on the loss of habitat and fragmentation of habitat which also forms the firm basis of barriers inflow of genes between two or more than two types of populations. A genetic pool is a complete set of all alleles investigated by the genetic material of all members of living species or a set of populations. A large pool indicates a greater level of diversity occurring in genetics that is associated with populations that survive as a result of selection phenomenon similarly low level of diversity leads to a reduced level of fitness and increase in chances of extinction of any species [27].

7.1 Process of genetic erosion

Bottlenecks of population results in the creation of genetic pools that possess very few mating partners which are fertile too. Reduction in the number of breed

Plants by unique genes will be similar to the situation where dealers operate with similar five cards again and again. Hence producing very few numbers of limited hands. As the sample inbreeds, it's both physical as well as reproductive effects appear to have existence much often. The most common and wide effects are on the Immune system which becomes weaker with time, presenting less resistance to diseases and in turn increases the count of bacteria, virus, parasites resulting in threats of diseases. So even if any endangered species in the genetic pool or bottleneck can bear with human development or growth. So, it faces the threat of epidemic which proves to be dangerous to the whole set of population [28].

7.2 Agricultural or crop plant hazardous loss from genetic erosion

Erosion in genes has resulted in the loss of a particular gene or gene which has undergone a recombination process i.e., complex set of genes that are either produced locally by the racers of land for domestic plants and animal species that adapt to natural conditions where these species grow. The major force behind the genetic erosion is the clearing of land, over the employment of species, deforestation, and degradation and finally grazing to a large extent. The major factors are the replacement of local varieties with varieties that are found to be non-locale. When commercial species overcome the traditional species and are introduced into the traditional farm system, it will also result in a reduction of a huge collection of varieties however the major problem is that it results in a reduction in tendency for uniformity in both genetic as well as economic factors in the development of a modern form of agriculture. So, if any endangered species can tolerate the process of human development and are adopted at a place much away from their natural habitat. It will still result in facing the danger of a serious threat to the whole population. With the advancement in science and technology, several techniques have been checked for checking defects of genetic erosion which result in the extinction of species that are at the extent of endangerment. But many techniques are very expensive for using them at a practical level. So, the best way is to preserve them by the protection of their natural habitat and to allow them to live in natural conditions as long as possible [29].

8. Endangered plant species and medicinal uses

Medicinal plants grow in. A natural environment around us and with the advent of technology, humans have gained the knowledge of how these can be utilized in fighting an illness or for maintenance of human health. The capability to use wild species in the improvement of health is not dependent on humans alone. The ability is affected by various factors such as pests, diseases, climate, environment, and other biotic or abiotic factors. According to society in America in 1999 the capability for maintainability of crop production depends on the compounds or genes being extracted from wild species of medicinal plants as depicted in **Table 1**. Because of their extensive use in commercial as well as the scientific environment, there has been increased pressure on wild species from which all these medicinal plants have been extracted. Over-harvest action, as well as commercial exploitation, has resulted in the unavailability of traditional medicine where the people utilize them. For all these reasons there is an urgent need to conserve these plant species [31].

Some of the examples of species include slippery elm which involves the use of gummy lining for used in North America as a therapeutic agent for cough and cold, gastrointestinal diseases, and allergy to skin epidermis. But this medicine being

Name of crop	Medicinal usage	Origin
Aloe Vera	For curing burns and wounds	South Africa
Aspirin	Pain killer, health rate normalization, and blood thinning	Europe
Bloodroot	Treatment of skin cancer	US
Camphor	As a pain-relieving agent	Asia
Digitalis	For the treatment of heart failure	Europe
Quinine	Malaria	South America.

Table 1.
Some medicinal plants which are at extent of endangerment [30].

used by local people is demanded by Millions of people now. They are not used commercially so the trees are separated from bark and are left to die. So, for about 50 pounds, 15 trees are sacrificed every year and now the species is identified as at risk by the US.

Another example is yew belonging to Texas specie that is used for the production of cancer drugs such as taxol. It is also identified as endangered based on its over-harvesting. Similarly, black cohosh is used for the cure of a large number of ailments such as colds, pains, and largely menopause and also found to be a list of endangerment due to degradation of habitat as well as over-harvesting. Moreover, goldenseal has numerous uses being used as a toner treatment of diseases such as hemorrhoids. It is already threatened, endangered, and vanished in many states.

9. Methodologies for the conservation of germplasm

There are various methods for the preservation of genetic resources or traits for crop plants. The easiest and most economical method is the storage of seeds of crop species. But not all the plant species can be stored easily in this condition because some seeds have a shorter span of viability and several species do not produce seed vegetative. Still there exist many methods that depend on storage conditions, storage vassals, the extent of conservation, and finally on the facilities available for proper storage of plants [32]. The most efficient way is to store biological crops in the environment where they were produced or developed earlier i.e., in situ farmlands as shown in **Figure 1**. This type of method can be employed when the natural environment is balanced and there are no chances of off balance. The most important point which must be considered is that genes should be conserved instead of the genotype of plant species [33].

9.1 In-situ conservation

This method employs the conservation of resources totally in a natural habitat. It involves the maintenance of plant species in which it grows and also in the habitat to which it is adopted for a long time in the past. The objective behind this methodology of preservation is to maintain the self-sustaining species in a natural ecosystem [34]. A huge collection of plant and animal species can be conserved by this mechanism. But along with it, there exists a limitation that it is impossible to sustain and preserve the genes of crops without conservation of the ecosystem of which it is adopted by nature [35]. It allows the conservancy of naturally occurring

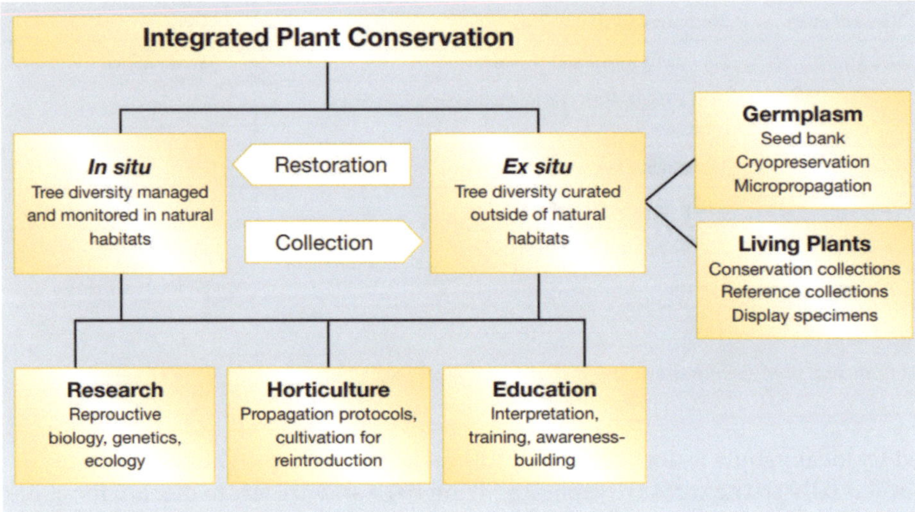

Figure 1.
Elaboration of integrated plant conservation involving both in-situ as well as ex-situ methodology. In-situ conservation allows the conservation of germplasm of and living parts of endangered plant species while ex-situ conservation involves the plant material available for research purpose, horticulture and reintroduction of materials preventing materials from getting extinct.

beneficial species in a condition where it continues to undergo evolution with time. Moreover, it also allows the conservation of both wild and cultivated genotypes without having much expenditure on the area. The major benefit of this process is that species selected by nature continue to evolve timely which results in the production of new recombinant forms of a living being. In the case of reluctant seeds that grow either in tropical or perennial regions, it serves as the best phenomenon of conservation within the in vitro environment [36]. Although in situ conservation is the best method but, its operation in any country or ecosystem is possible only when it is ensured by people who are in continual conflict with national plans and when its continuity is also confirmed in any environment. Usually, if continuous and control monitoring does not operate, its survival rates get much lower. This results in loss of naturally occurring habitat and also replacement of old generation of crops with new varities which occur as a normal part of the crop growing system [37].

In situ conservation is, therefore, can be achieved by the protection of naturally occurring wild species in their natural or adopted habitat via cultivation in fields. Such areas or regions are being discovered in natural parks and recreational areas under government territory. Practices such as horticulture and floriculture present an efficient way of conservation in the naturally existing ecosystem [38]. Horticulture is a phenomenon in agriculture whereby plants are preserved for the purpose of feed but mainly for comfort and decoration purposes. It employed the use of knowledge and skills to grow plants for use in both food and non-food areas and also for social requirements. It includes both tropical and perennial species, vegetable varieties, tasty fruits in addition to decorative indoors, and other Landscape plant varieties [39]. Floriculture is also a subtype of horticulture which is mainly concerned with cultivating flower varieties of ornamental plants for use in the floral industry, gardens, and orchids. Development and growth of varieties via breeding techniques to a novel variety of species is a major point of focus in floriculture which allows the transfer of desirable characteristics to next-generation resulting in maintenance of specific genetic traits [40]. This methodology of preservation has some advantages which include;

- Each preserved area will contain a small portion of whole diversity i.e., Small portion of total diversity. So, it requires the preservation of a large number of areas for the conservation of the whole genetic pool.

- The maintenance and management of all these areas also require labor and present problems.

- This is the most expensive method for the conservation of germplasm.

9.2 Ex-situ conservation

Another methodology involving the collection of plant gardens and banks of seeds where the plants are grown under natural conditions [41]. Seed banks are maintained and produced by research institutions and universities produced via the technique of tissue culture and utilization of much lower temperature in the environment for its operation [42]. It also ensures that the plant materials are easily available, characterization is done efficiently and well documented and its exposure to the outer environment is safe i.e., it should not possess any threat to the natural ecosystem of humans as well as animals. However, it provides the best alternative to naturally existing methods which in addition to providing an opportunity to wild species having desirable traits to continue undergoing the evolutionary process in a naturally existing environment. This method has the advantage of safeguarding the germplasm while it is in its natural environment resulting in the genetic variation in naturally occurring verities and is readily available for use. Examples of plant undergoing this subtype of preservation under biotechnological area are sugarcane, cocoa, and maize, etc. [43].

This is also referred to as offsite conservation which employs the conservation of species outside their natural habitat or system. In this method, the genetic information of the plant is preserved in form of banks which may be either seed or gene bank or in the form of cultures to increase their half-life so that they can be used for a long period inefficiently [44]. The class of preservation technique results in the formation of collection or bank of genes, DNA, seeds, and germplasm forming a genetic library in the form of gardens. This will lead to the creation of a good option for the conservation of species that are thought to be endangered or near the extent of it, which are primitive and in turn, are much valuable for use in industry for commercial purposes. It includes certain techniques such as cryopreservation and other genetic transfer approaches for the eradication of diseases, pest and stress control, and lastly conservation of endangered species in the long run [45]. It is almost similar to that of in-vitro methodological practice. Other disadvantages include loss of viability of seed structure, destruction of the crop by pest or insect, poor germination of a seed plant, and lastly, it is a much expensive procedure [46]. On the other hand, major advantages can be summarized as,

- Small areas can store a large collection of materials

- It protects all other environmental-based methods [47].

- It is the cheapest method and preservation of germplasm is much easier.

- It is possible to store the whole genetic material in a single place.

The most advanced form of preservation of genetic resources is to maintain them in laboratory conditions. This is the conservation technique which employs the use of test tubes or laboratory apparatus which is sealed in one or other way for maintenance of resources [48]. The genetic resources such as tissue cells or callus are placed in the sealed tubes which operate on the fact that plant parts can be kept alive under controlled laboratory conditions which proves the fact that plants are totipotent. This means that every part of a plant can develops into a whole organism. This phenomenon has made this fact clear that disease-free plants and species can be transferred to the next generation within the laboratory controlled conditions [49]. Or in other words, engineered species provide a viable means for the transfer of pest and insect-free species from one generation to another. The source of such genotypes is from the culture of laboratories or having origins from international seed banks [50].

In vitro conservation of plants was first done in the mid-1970s. Although whole 'is not can be regenerated from any part of the plant because of its totipotency but due to the involvement of unorganized culture there exist some risks of a generation of somatic mutation and mutants. In comparison to it, the cultures containing somatic meristem culture are much more stable in their transformation mode but also it can propagate more frequently as these areas do not have to recover after differentiation [49]. Most efficient storing systems are usually not much expensive, are easy to maintain, and reduce the work labor and load in germplasm working bank. Scheduled monitoring of the cells along with viability and contamination assessment is not that necessary. The exploitation of in Vito technique of genetic conservation is hindered if any species is unable to prorogate to the next generation from tissue or cell culture [51]. For example, a proper technique for the prorogation of coconut does not exist yet unlike other crops of this class which can be propagated inefficient way via callus differentiation. But in this case, the leaves or plantlets can only be produced from a zygote or embryo. Each embryo in this case will produce a new plant which represents no further division of genetic material. In the same manner, the effect of in-vitro culturing is much less for woody plants as compared to other species as it can result in difficulty in culturing and regeneration of new species [52]. In these cases, less research has been done for the development of an appropriate cultural technique in vitro. But a thorough examination of the problem occurring in the handicap pathway of procedure for wood culture can solve this problem, presenting a suitable solution to the development of plant and conservation of their genetic resources efficiently [53].

The most important drawback of this phenomenon is that it requires the utilization of modern technology and labor force under the controlled conditions of an aseptic environment. Also, it requires proper laboratory skills with excessive usage of electricity which makes this procedure much labor-intensive and expensive [54]. This process is helpful in the production of disease-free varities of plants that are also pest-free and these species include sugarcane etc. The produced genetic resources are used in several ways such as genetic improvement, maintenance of biodiversity, mechanism-based research of ecosystems, classification according to taxonomy, monitoring of environmental characteristics, epidemiological, and forensic based studies. One of the main strategic reasons behind germplasm conservation is that it maintains biological diversity and provides germplasm which is validated in both genotypic and phenotypic aspects [55]. Germplasm is either conserved in the form of seed or meristem form.

10. Gene bank

Gene banks are the type of repository in biology for the preservation of genetic resources. In the case of plants, it is done by storing in laboratory conditions, freezing cuts from the plant materials, or maintaining stocks of seed. Accession is the term provided to each sample in a gene bank like the species or variety. In plants, it is easy to unfreeze the materials for their propagation and usage.

Gene banks are also classified as both in vivo that is within the body and in-vitro which involve sustaining of characters in proper laboratory conditions. The type of gene bank where traits or alleles are stored by employing conventional methodologies are termed in-vivo. For example in the form of seeds and vegetative collections [56]. While the subtypes where the characteristic resources are stored in form of non-conventional methods in form of cellular structure and tissues are referred to as in vitro. Both techniques are of prime importance in the development of valuable trait crops for breeders to develop both new and improved varieties [57]. This involves using DNA as a source of DNA in terms of germplasm being employed in breeding technologies. When these are properly identified and after that efficiently characterized, it will result in the production of the transgenic organism which can express these genes. Genetic disruption can be avoided by the phenomenon of transformation which also involves sexual hybridization. It is not limited by compatibility from the sexual life cycle and can be evolved from other forms of life in the short run. The transgenic genes are helpful in the production of plants which in addition to herbicide-resistant are also pest resistant and are conveniently preserved as a transgenic or cloned form of genetic material. The process is limited by the identification of potable genes which will result in the production of higher yield along with greater stability of Transformants in host genera. Such genes have been produced successfully for conservation as well as patent purposes so that they can be employed at a commercial level. Economically or technically, is not worthy in the future that this synthesis of the gene will store these genetic traits in the form of physical germplasm i.e., in form of seeds, tissue, or the whole plant, etc. Conservation of DNA molecules and similarly the assembly of these molecules in the form of DNA data sequence is not the best alternative to conventional methods for germplasm because genes are not coordinated in them in a small similar fashion.

Recently with the discovery of artificial chromosomes in yeast has raised the fact that coordinated assemblies of genes can be made and therefore can be conserved which will allow further morphological or phenotypical changes to be engineered in the laboratory efficiently. To use them practically, it is important to conserve the host organism but the genetic information in them is not yet fully discovered. But gene liberates, sequence data, and gene banks cannot be employed to reproduce a whole organism but have a significant role in preserving genetic resources of crop plants which are either on the extent of danger or found to be endangered. The conservation of plant or genetic diversity involves the collection of small parts of plants such as tissues, cells, shoots, etc. A tissue sample from all the plants and species will be collected in liquid nitrogen as described in cryopreservation at a much lower temperature of freezing point. In theory, these samples are not indefinite and DNA extraction is not performed until recommended. So, at that moment, DNA can be identified, immobilized in a membrane to act as a source of a specific gene or sequence of DNA molecules. The technique is helpful in the conservation of both undefined and undescribed species of plants whose seeds cannot be stored directly and is used for diverting those whose seed values have been found earlier and observed in germplasm banks already. DNA sequencing is carried out in almost all laboratories throughout the world because they can be compared to novel sequences with those which are properly and characterized considerably. Comparisons also highlight the unrevealed homologies and suspected functional properties between the organisms which are unrelated. Most organizations support coordinated DNA sequencing and storage. The most famous banks are the European molecular biology laboratory located in the US, GenBank operated by the United States laboratory.

The rapidly increasing data of sequence raises voice on important problems of storing and facilitating rapid comparison regarding new information on data sequence of the gene. Gene banks are of three types which include;

10.1 Seed bank

It stores the seed at very low temperatures after they have been dried efficiently. Spores, as well as meristems, are stored by this method in seed banks but the vegetative plants which do not possess seed are not concerned by this technique. The largest bank for storing seeds is Millenium seed bank which is located at WTMB near London.

10.2 Tissue bank

By this technique buds, roots and meristem are stored in this way by utilizing light, the temperature in presence of aseptic conditions, and nutrient media containing all the essentials such as carbon, nitrogen, etc. Hence the technique is based on the preservation of seedless plants and which reproduce by sexual reproduction.

10.3 Cryobank

By this technique, a seed or embryo cell is stored at a much lower temperature or temperature much lower than freezing temperature in liquid nitrogen at a temperature of about −196 °C. So, it is helpful for the conservation of species that are at the extent of extinction or have become endangered. But mostly it is used for the cryopreservation of genetic resources of animals. But in the case of plants, conservation of pollen grains is done by storing at a much lower temperature of −196 °C. Hence this methodology is used for cross breeding and also in the production of plants having a set of chromosomes.

10.4 Field gene bank

The method is used for the conservation of genes of planting plants where the ecosystem is created artificially. By this method, the different plant species can be compared so that they can be studied in detail. It needs more artificial requirements such as soil, water, and weather, etc. The germplasm of most crop plants is conserved by this process. For example, 43000 rice varieties are conserved in Orissa at a central research institute.

11. Cryopreservation

The word is derived from two Greek words. i.e., Kryos means frosting while preservation means storage for a long time or increasing half-life in one way or another. Following these techniques cells and tissues are stored at a much lower or frozen temperature either using carbon dioxide at −79 °C or nitrogen gas at −160 °C in the form of vapors in deep freezers. In the c as e of liquid nitrogen, the limit of temperature would be from 170 °C to 197 °C. The technique involves four stages involving freezing, thawing, and re-culturing, etc. [58]. Thus, freezing temperature inactivates the cells and tissues so that it can be preserved for a longer period. Any of the tissues of the plant can be preserved under proper conditions for example meristem, stem, ovules, anther, embryos, endosperm, cells, and leaves, etc. [59].

The process of cryopreservation being followed by regeneration of the whole plant invoke the following steps **Figure 2**;

- Isolation and development of sterile tissue culture
- Addition of cryoprotectants
- Pretreatment
- Freezing of plant material
- Storing of plants parts
- Thawing followed by culturing
- Assessment of viability of cells based on their rate of survival
- Regeneration of plants

11.1 Isolation of sterile tissue or cell

Physiological and structural conditions of plant effects ultimately the survival of the plant during cryopreservation. Tissues to be used in preservation must be healthy, small, young, having rich cytoplasm, and highly vacuolated. In either case, callus acts as the best source of tissues as it is more resistant to damage caused by freezing [60]. So, a callus after 1 to 2 weeks of subculturing is selected for the

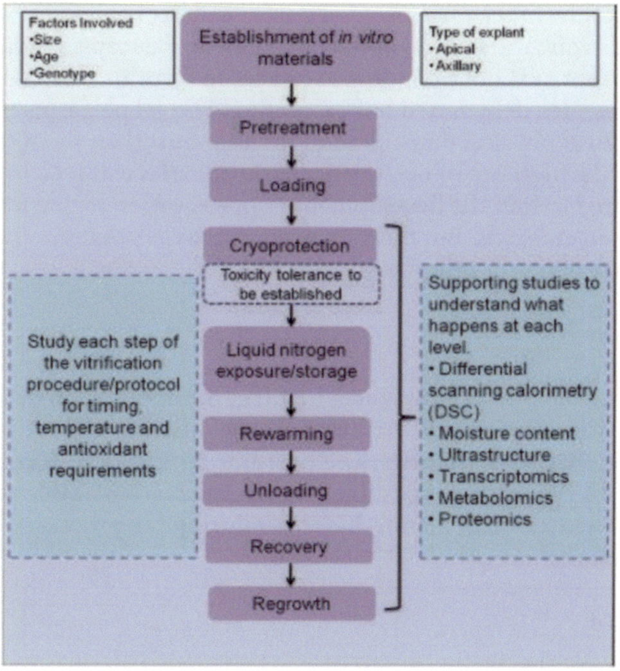

Figure 2.
Schematic flow sheet representation of steps involved in cryopreservation of plant materials in biotechnology for genetic resource(germplasm) conservation of endangered plant species such as golden paintbrush from natural environment to cryobank in form of seeds, tissues, roots, meristem and shoots etc.

cryopreservation process. But old and black areas should be avoided and organized structures are preferred more.

11.2 Addition of cryoprotectants

To prevent the damage resulting from abrupt freezing or thawing, the chemicals such as glycerol, alcohol, dimethyl sulfate, glycerol, praline, etc. are being added for the purpose of conservation. This protectant referred to as cryoprotectants are added to protect freeze cells or low freezing temperatures etc. But the limitation of the procedure is that only a few biological materials can be frozen below the minus temperature in presence of gas without affecting the viability of cell structures. Liquid nitrogen is used because of the following reasons;

- It is inert in chemical form.
- It is a less expensive process posing less burden on the economy.
- It is non-toxic posing no side effects on the environment.
- It is a non-flammable and most readily available method so far.

11.3 Vitrification

However, two more practical approaches in biotechnology may lead to widespread applications of conservation of germplasm of plants to reduce damage from abrupt cooling. This includes vitrification by using cryoprotectants mixture and another is an encapsulation of a sample with gel which is dehydrated later on as described above. For the process of vitrification, the sample is submerged in a cryoprotectants mixture which results in the promotion of conservation of cellular water into non-crystal-like solids which later cool rapidly. In the case of encapsulation, the sample material such as root or shoot tip is dipped in a gel to form an artificial seed-like structure which is then dehydrated before cooling. The gel performs the function of protection against physical damage and is more robust than shoot tip or embryo culture. Despite the presence of optimistic methodologies using plant tissues, further research is required to find the development of preservation to that available for animal and human embryos. But there exist many barriers that prevent the utilization of technology in one or other way.

11.4 Pretreatment

The process involves regrowth which involves the application of additives to enhance growth e.g. abscisic acid etc. [61]. On the other hand, cryoprotectants act as an anti-freeze, increases viscosity, and prevents damage which resulted wither due to the formation of ice crystals during cryopreservation or due to an increase in intracellular concentration of solutes before or during the process of freezing as a result of dehydration.

11.5 Dehydration

Vitrification is a process of conversion of liquid into solid in the absence of crystallization. When the cells have properly undergone the process of slow freezing, it will result in vitrification where ice formation does not take place because here the aqueous solution is much concentrated which results in permitting the formation of ice cubes. Instead, the water gets solidified into a glassy clear state. Dehydration during this process is achieved by the high concentration of osmotically active compounds like sugars, polyols performed in the sterile cabinet over silica gel [62]. In the process of dehydration, a reduction in the amount of water followed by the formation of ice and an increase in osmotic pressure occurs which ultimately depress the freezing point [63].

11.6 Rapid freezing

Then plant material is placed in liquid nitrogen at a much lower temperature ranging from −300 to −1000 °C. Dry ice can also be similarly used in the process. The more quickly freezing is done the less will be the intracellular state of crystal formation. The methodology is simpler and easy to handle and can also be used for tips of potatoes and strawberry species. Dry ice can also be utilized for this purpose.

11.7 Stepwise freezing method

In this method, the temperature is lowered to about −30 °C for at least a period of 30 minutes and then abrupt cooling is done via using liquid nitrogen at the much lower temperature of −196 °C. Slow freezing increases dehydration while abrupt freezing promotes crystal formation. It gives excellent results in the preservation of strawberries in the suspension culture.

11.8 Storage

Storing at the correct temperature is as important as that of freezing. For storage, the temperature is left to almost −70 °C to 197 °C because this temperature is sufficiently low for the preservation of cells without metabolic damage to them. Long term storage is mostly done at about −197 °C.

11.9 Thawing

It involves the rapid thawing of the ampule containing the sample in a water bath at about 40 °C. They are plunged into warm water with swirling for rapid mixing just to the point where ice gets disappears and is important for the survival of tissues that the sample must be removed from the water bath after melting of ice. Tissues being thawed at a much lower temperature are then abruptly thawed following this step.

11.10 Determination of rate of survival

Regrowth of stored tissue is the best indication of the survival of plant tissues. For this purpose, many viability tests are used which involve fluorescein diacetate staining, measurement of growth by cell number, and finally by calculating the dry and fresh weight. The two most popular methods are m

- Triphenyl tetrazolium chloride
- Evans blue staining

It provides the best opportunity for the conservation of endangered species being used for medicinal purposes. It also provides an ideal approach for the suppression of cell division to avoid further need for sub culturing. Pathogen free cultures and subcultures can be frozen and stirred when required and also provide a much suitable material for the selection of cold-resistant strains of mutant cell lines which later get differentiated into frost resistant cells or plants. The seeds may be stored for food crops or to protect biodiversity or reason for storage also varied which involve first drying of moisture to less than 5% and then stored at the much lower temperature of −18 °C or below it [64].

12. Work done by the vavolian center of biotechnology

For the last 20 years, advancement in tissue culture technologies has led to the development of the micro propagation method which is a novel technology, providing young and fresh plants for horticulture, agricultural, and agriculture purposes. One of the main consequences is the rapid growth of in-Vito exchange as a viable means for the transfer of germplasm between different laboratories [65]. The International Board for genetic research has elucidated that more than 140 plant germplasm has been exchanged from 1980 to 1986s. Out of all attempts, about 97% were found to be successful. Now almost every agriculture research center is attributed to the exchange of germplasm in-vitro. For example, the exchange of germplasm of potato culture is now a routine procedure. Shoot cultures are incubated for about 3 weeks after inoculation to induce roots and any contamination resulting from the microbial mass [66]. Transferring them to a fresh medium compensates this problem and increases rates of survival. Now the new era has replaced the cultures with small test tubes produced under in-vitro conditions [67]. These are more robust and rapid methods, and the produced plants or germplasm can be stored for months or even years. The recipient can place them in nursery beds without the involvement of further culture step. In the modern process of potato breeding, the in-vitro cultures provide another species that is disease free followed during the process of field testing required to select the most desirable form of clone as in **Table 2** [69].

The successful application and conservation of genetic resources in every country for the purpose of food and agriculture depend upon the collaboration of the government. Policymakers, germplasm scientists, rural populations, and

Species/crops	Applications
Rice	Food, fodder, and beverage
Sorghum	fodder
Cowpea	Food, fodder
Maize	Food, industry, fodder
Soya bean	Industry, food, fodder
Sesame	fodder
Cassava	fodder
Millet	Medicinal, fodder, industry
Hungry rice	Food, industrial, fodder
Yarn	Food
Sugarcane	Food, beverage, industry
Groundnut	food

Table 2.
Utilization of plant species/crops in various aspects after their conservation in the form of germplasm which are thought to be endangered in future [68].

breeders or farmers. Usually plant genetic resources are conserved because they are ultimately used in food and agriculture and sustainable agriculture depends upon their usage. Farmers in modern agriculture use their plants or crops for purpose of not only food but for medicine and fodder also as shown in **Table 3**. Deployment of genetic resources in a better idea paves the way for the reduction of vulnerability of crops or plants to that of insects, pests, and other fertilizers making them herbicide, pesticide, and insecticide resistant. In national research institutes of every country about 13% rice, 7% soybean, and 8% of sugarcane species are conserved in various breeding technologies as illustrated in **Table 1**. With the abrupt increase in population and reduction in land available for agricultural purposes, an increment in the production of food, as well as its distribution across the globe, is much necessary. There is an utmost need in every country to use their genetic resources for better purposes utilizing the breeding techniques effectively. The involvement of genetic resources in the techniques has resulted in almost the compensation of food required by the increasing population of the world. So the stress of poverty alleviation in developing countries has been reduced effectively and is also involved in food security depending on the availability and utilization of species that produce a higher amount of crops or plants with desirable characters especially in rural areas where most families rely on farming for their survival. So utilizing a small collection of gene bank resources can lead to greater benefits as elaborated in breeding programs. However, less usage of them can lead to fewer benefits in both the social and economic sectors. The constraints involved in the low level of germplasm conservation include lack of ability to characterize and evaluate gene data banks, insufficient knowledge, inappropriate documentation and poor relationship between users of gene banks and germplasm. Currently, in sugarcane industry much data is available on characterization and evaluation of data on sugarcanes for utilization by stakeholders.

Name of crop	Type of conservation	Improvements resulted
Soybeans	Gene banks, bottles, and tissue culturing	Reducing days required for maturation. Transferring smut resistance from wild to cultivated type.
Sugarcane	Gene banks	Transfer of smut resistance from wild to cultivated type. Introduction of sucrose and improved protein content.
Rice	Storing bottles and tissue culturing	Hybridization of both wild and cultivated species. Development of short duration techniques Development of iron resistant specie.
Sorghum	Field gene banks, tissue culture	Production of short duration variety. Mildew resistant varieties.
Maize	Storing bottles and tissue culturing	Production of varieties which are high in lysine content
Cowpea	Storing bottles and tissue culturing	High yield of product. High protein species.
Sesame	Storing bottles and tissue culturing	Early maturing species. Development of verities that are resistant to black specks. High oil containing verities.

Table 3.
Improvement in crops by germplasm conservation of plants (conservation type) being used in everyday life which are thought to get endangered in future [70].

13. Conclusion

The development of various successful methodologies for preservation of genetic resources enables the establishment of basal collections of endangered species with the representative diversity. The collections include many of species which have been threatened for loss of habitat. Collections related to critically endangered species are maintained in simple media. These threatened species could also be maintained at lower temperature after acquiring specific laboratory conditions. From the whole discussion, the fact has become clear that there is a need to conserve germplasm in a variety of ways. It is also elucidated that the effectiveness of conservation technology depends upon the maintainability of the collection of genetic resources in a much cost-effective manner. So, there is an abrupt need to emphasize evaluation of the efficiency of conservation procedure by realizing efforts in an in-expensive way. Because the future will depend upon the presence and utilization of conserved germplasm. Thus, there is an urgent need to conserve the most important crops such as maize, rice, sorghum, wheat, etc. The variation in them is leading to genetic erosion which will ultimately degrade them and their availability in the future will be affected. So on the whole, involvement of germplasm in its preserved form and plant breeders in growth and development of these crops for improved and better varieties. Moreover there is a large collection of endangered species in under developing countries and creation of botanical gardens for in vitro and in vivo conservation allows establishment of culture facilities which in turn results in rescue as well as regrowth of endangered species. Additionally networking in exchange of information or materials and dissemination of useful protocols are important steps in continuous development and exchange of mechanisms for germplasm conservation. So at this point scientists and conservationists need to work together in order to develop better programs for germplasm conservation of plant species which are at risk of extinction.

Author details

Sameer Quazi[1,2*], Tanya Golani[3,4] and Arnaud Martino Capuzzo[5,6]

1 Department of Genetics, Indian Academy Degree College, Bangalore, Karnataka, India

2 Founder and CEO, GenLab Biosolutions Private Limited, Bangalore, Karnataka, India

3 Department of Medical Genetics, University of Glasgow, Scotland, United Kingdom

4 Senior Research Associate, Department of Genetics and Molecular Sciences, GenLab Biosolutions Private Limited, Bangalore, Karnataka, India

5 Department of Biological Sciences, University of Milan, Italy

6 Senior Research Associate, Department of Life Sciences, GenLab Biosolutions Private Limited Bangalore, Karnataka, India

*Address all correspondence to: colonel.quazi@gmail.com

References

[1] Engelmann, F. (1997). *In vitro germplasm conservation.* Paper presented at the International Symposium on Biotechnology of Tropical and Subtropical Species Part 2 461.

[2] Bonner, F. T. (1990). Storage of seeds: potential and limitations for germplasm conservation. Forest ecology and management, *35*(1-2), 35-43.

[3] Plucknett, D. L., Smith, N., Williams, J., & Anishetty, N. M. (1983). Crop germplasm conservation and developing countries. Science, *220*(4593), 163-169.

[4] Scowcroft, W. R. (1984). Genetic variability in tissue culture: impact on germplasm conservation and utilization.

[5] Frankel, O. (1984). Genetic perspectives of germplasm conservation. Genetic manipulation: impact on man and society.

[6] Singh, R., Gautam, P., Saxena, S., & Singh, S. (2000). Scented rice germplasm: conservation, evaluation and utilization. *Aromatic rices. Kalyani, New Delhi*, 107-133.

[7] Bretting, P. K., & Duvick, D. (1997). Dynamic conservation of plant genetic resources. Advances in Agronomy, *61*(1), 51.

[8] Lambardi, M., & De Carlo, A. (2003). Application of tissue culture to the germplasm conservation of temperate broad-leaf trees. In *Micropropagation of woody trees and fruits* (pp. 815-840): Springer.

[9] Villalobos, V. M., Ferreira, P., & Mora, A. (1991). The use of biotechnology in the conservation of tropical germplasm. Biotechnology advances, *9*(2), 197-215.

[10] Verma, N., Mohanty, A., & Lal, A. (2010). Pomegranate genetic resources and germplasm conservation: a review. *Fruit, Vegetable and Cereal Science and* Biotechnology, *4*(2), 120-125.

[11] Greene, S. L., & Hart, T. C. (1999). Implementing geographic analysis in germplasm conservation. *Linking genetic resources and geography: emerging strategies for conserving and using crop* biodiversity, *27*, 25-38.

[12] Balogun, M. O. (2009). Microtubers in yam germplasm conservation and propagation: The status, the prospects and the constraints. *Biotechnology and Molecular Biology Reviews, 4*(1), 1-10.

[13] Fay, M. F. (1992). Conservation of rare and endangered plants using in vitro methods. *In* Vitro Cellular & Developmental Biology-Plant, *28*(1), 1-4.

[14] Waples, R. S. (2002). Definition and estimation of effective population size in the conservation of endangered species. *Population viability analysis*, 147-168.

[15] Channell, R., & Lomolino, M. V. (2000). Dynamic biogeography and conservation of endangered species. Nature, *403*(6765), 84-86.

[16] Haydon, D., Randall, D., Matthews, L., Knobel, D., Tallents, L., Gravenor, M., Woolhouse, M. (2006). Low-coverage vaccination strategies for the conservation of endangered species. Nature, *443*(7112), 692-695.

[17] Langpap, C. (2006). Conservation of endangered species: Can incentives work for private landowners? Ecological economics, *57*(4), 558-572.

[18] Wade, R., Augyte, S., Harden, M., Nuzhdin, S., Yarish, C., & Alberto, F. (2020). Macroalgal germplasm banking for conservation, food security, and industry. *PLoS biology, 18*(2), e3000641.

[19] Carra, A., Carimi, F., Bettoni, J. C., & Pathirana, R. (2019). Progress and Challenges in the Application of Synthetic Seed Technology for Ex Situ Germplasm Conservation in Grapevine (Vitis spp.). In *Synthetic Seeds* (pp. 439-467): Springer.

[20] Shahzad, A., Parveen, S., Sharma, S., Shaheen, A., Saeed, T., Yadav, V., Upadhyay, A. (2017). Plant tissue culture: applications in plant improvement and conservation. In *Plant Biotechnology: principles and applications* (pp. 37-72): Springer.

[21] Surendran, K., Nair, R. A., & Pillai, P. P. (2020). Molecular Markers and Their Application in the Identification of Elite Germplasm. In *Plant Metabolites: Methods, Applications and Prospects* (pp. 57-70): Springer.

[22] Crisci, J. V., Sala, O. E., Katinas, L., & Posadas, P. (2006). Bridging historical and ecological approaches in biogeography. Australian Systematic Botany, *19*(1), 1-10.

[23] Olomola, D., Aguda, S., Olorode, E., Oyediran, R., & Adekunle, E. (2019). The application of biotechnology in biodiversity conservation.

[24] Cuevas, H. E., Rosa-Valentin, G., Hayes, C. M., Rooney, W. L., & Hoffmann, L. (2017). Genomic characterization of a core set of the USDA-NPGS Ethiopian sorghum germplasm collection: implications for germplasm conservation, evaluation, and utilization in crop improvement. *BMC genomics, 18*(1), 1-17.

[25] Scott, J. M. (2001). Endangered species and peripheral populations: cause for conservation. Endangered Species Update, *18*(5).

[26] Shinde, R. (2020). Introduction to transgenic animals and their applications, Germplasm storage and cryopreservation.

[27] Comizzoli, P., & Holt, W. V. (2014). Recent advances and prospects in germplasm preservation of rare and endangered species. In *Reproductive sciences in animal conservation* (pp. 331-356): Springer.

[28] Admas, S., Tesfaye, K., & Haileselassie, T. (2018). Application of advanced genomics for conservation and utilization of plant genetic resources. *J. Environ. Agric. Sci, 17*, 30-41.

[29] Faisal, M., & Alatar, A. A. (2019). *Synthetic seeds: germplasm regeneration, preservation and prospects*: Springer Nature.

[30] Goodrich, J., & Buskirk, S. (1995). Control of abundant native vertebrates for conservation of endangered species. Conservation Biology, *9*(6), 1357-1364.

[31] Reyes-Valdés, M. H., Burgueño, J., Singh, S., Martínez, O., & Sansaloni, C. P. (2018). An informational view of accession rarity and allele specificity in germplasm banks for management and conservation. *PloS one, 13*(2), e0193346.

[32] Muthoni, J., Mbiyu, M. W., & Nyamongo, D. (2010). A review of potato seed systems and germplasm conservation in Kenya. Journal of Agricultural & Food Information, *11*(2), 157-167.

[33] Reed, B. M., Dumet, D., Denoma, J. M., & Benson, E. E. (2001). Validation of cryopreservation protocols for plant germplasm conservation: a pilot study using Ribes L. Biodiversity & Conservation, *10*(6), 939-949.

[34] Bockelman, H. E., Valkoun, J., & Ullrich, S. (2010). Barley germplasm conservation and resources. *Barley: improvement, production, and uses. Wiley-Blackwell, Oxford, UK*, 144-159.

[35] Dubale, P., & Teketay, D. (2000). *The need for forest coffee germplasm*

conservation in Ethiopia and its significance in the control of coffee diseases. Paper presented at the Proceedings of Coffee Berry Disease Workshop. Addis Ababa, Ethiopia.

[36] Berding, N., & Koike, H. (1980). Germplasm conservation of the Saccharum complex: A collection from the Indonesian Archipelago. Hawaiian Planters' Record, *59*(7), 87-178.

[37] Swamy, M. K., Balasubramanya, S., & Anuradha, M. (2009). Germplasm conservation of patchouli (Pogostemon cablin Benth.) by encapsulation of in vitro derived nodal segments. International Journal of Biodiversity and Conservation, *1*(8), 224-230.

[38] Gupta, P., & Varshney, R. K. (1999). Molecular markers for genetic fidelity during micropropagation and germplasm conservation? Current Science, *76*(10), 1308-1310.

[39] de Vicente, M. C., Guzmán, F. A., Engels, J., & Rao, V. (2006). Genetic characterization and its use in decision-making for the conservation of crop germplasm. *The role of biotechnology in exploring and protecting agricultural genetic* resources, *129*.

[40] Tay, D. (2007). Herbaceous ornamental plant germplasm conservation and use. In *Flower breeding and genetics* (pp. 113-175): Springer.

[41] LU, X.-x., & CHEN, X.-l. (2003). Progress of Conservation and Research of Crop Germplasm Resources in China [J]. Scientia Agricultura Sinica, *10*.

[42] Maruyama, E., Kinoshita, I., Ishii, K., Ohba, K., & Saito, A. (1997). Germplasm conservation of the tropical forest trees, Cedrela odorata L., Guazuma crinita Mart., andJacaranda mimosaefolia D. Don., by shoot tip encapsulation in calcium-alginate and storage at 12-25° C. Plant Cell Reports, *16*(6), 393-396.

[43] González-Benito, M., Ramírez, I. C., & Aranda, J. M. L. (2004). the use of cryoperservation for germplasm conservation of vegetatively propagated crops. *Spanish Journal of Agricultural Research*(3), 341-352.

[44] Malice, M., & Baudoin, J.-P. (2009). Genetic diversity and germplasm conservation of three minor Andean tuber crop species. Biotechnologie, Agronomie, Société et Environnement, *13*(3), 441-448.

[45] Rong, J., Xia, H., Zhu, Y., Wang, Y., & Lu, B. R. (2004). Asymmetric gene flow between traditional and hybrid rice varieties (Oryza sativa) indicated by nuclear simple sequence repeats and implications for germplasm conservation. New Phytologist, *163*(2), 439-445.

[46] Engelmann, F. (2012). Germplasm collection, storage, and conservation. In *Plant biotechnology and agriculture* (pp. 255-267): Elsevier.

[47] LIU, Y., QIU, Y. P., ZHANG, L., & CHEN, J. (2005). Dormancy breaking and storage behavior of Garcinia cowa Roxb.(Guttiferae) seeds: implications for ecological function and germplasm conservation. Journal of Integrative Plant Biology, *47*(1), 38-49.

[48] Heraty, J. M., & Ellstrand, N. C. (2016). Maize germplasm conservation in Southern California's urban gardens: Introduced diversity beyond ex situ and in situ management. Economic Botany, *70*(1), 37-48.

[49] Demissie, A., & Bjørnstad, Å. (1997). Geographical, altitude and agro-ecological differentiation of isozyme and hordein genotypes of landrace barleys from Ethiopia: implications to germplasm conservation. Genetic Resources and Crop Evolution, *44*(1), 43-55.

[50] Brush, S. B. (1991). A farmer-based approach to conserving crop

germplasm. Economic Botany, *45*(2), 153-165.

[51] Di Guardo, M., Scollo, F., Ninot, A., Rovira, M., Hermoso, J., Distefano, G., . . . Batlle, I. (2019). Genetic structure analysis and selection of a core collection for carob tree germplasm conservation and management. *Tree Genetics & Genomes, 15*(3), 41.

[52] Cyr, D. (2000). *Cryopreservation: roles in clonal propagation and germplasm conservation of conifers.* Paper presented at the Cryopreservation of tropical plant germplasm: current research progress and application. Proceedings of an international workshop, Tsukuba, Japan, October, 1998.

[53] Ji, P., Li, H., Gao, L.-Z., Zhang, J., Cheng, Z., & Huang, X. (2011). ISSR diversity and genetic differentiation of ancient tea (*Camellia sinensis var. assamica*) plantations from China: implications for precious tea germplasm conservation.

[54] Lawrence, M., Marshall, D., & Davies, P. (1995). Genetics of genetic conservation. I. Sample size when collecting germplasm. Euphytica, *84*(2), 89-99.

[55] Hartati, S., Muliawati, E. S., Pardono, P., Cahyono, O., & Yuliyanto, P. (2019). Morphological characterization of Coelogyne spp for germplasm conservation of orchids. Revista Ceres, *66*(4), 265-270.

[56] Heywood, V. H. (1991). Developing a strategy for germplasm conservation in botanic gardens. *Tropical Botanic Gardens: Their Role in Conservation and* Development, 11-23.

[57] Bajaj, Y. (1995). Cryopreservation of plant cell, tissue, and organ culture for the conservation of germplasm and biodiversity. In *Cryopreservation of plant germplasm I* (pp. 3-28): Springer.

[58] Colunga-GarcíaMarín, P., & Zizumbo-Villarreal, D. (2006). Tequila and other Agave spirits from west-central Mexico: current germplasm diversity, conservation and origin. In *Plant Conservation and Biodiversity* (pp. 79-93): Springer.

[59] Pengelly, B. C., & Maass, B. L. (2019). Tropical and subtropical forage germplasm conservation and science on their deathbed! 2. Genebanks, FAO and donors must take urgent steps to overcome the crisis. Outlook on Agriculture, *48*(3), 210-219.

[60] Leslie, C., Seybold, S., Graves, A., Cranshaw, W., & Tisserat, N. (2009). *Potential impacts of thousand cankers disease on commercial walnut production and walnut germplasm conservation.* Paper presented at the VI International Walnut Symposium 861.

[61] Vuylsteke, D. (1998). *Shoot-tip culture for the propagation, conservation and distribution of Musa germplasm*: IITA.

[62] Gopal, J., Chamail, A., & Sarkar, D. (2004). In vitro production of microtubers for conservation of potato germplasm: effect of genotype, abscisic acid, and sucrose. *In* Vitro Cellular & Developmental Biology-Plant, *40*(5), 485-490.

[63] Liu, C.-Z., Murch, S. J., Jain, J. C., & Saxena, P. K. (2004). Goldenseal (Hydrastis canadensis L.): in vitro regeneration for germplasm conservation and elimination of heavy metal contamination. In Vitro Cellular & Developmental Biology-Plant, *40*(1), 75-79.

[64] Potgieter, J., & Mashope, B. (2007). *Cactus pear (Opuntia spp.) germplasm conservation in South Africa.* Paper presented at the VI International Congress on Cactus Pear and Cochineal 811.

[65] Yan, W., Rutger, J. N., Bryant, R. J., Bockelman, H. E., Fjellstrom, R. G., Chen, M.-H., . . . McClung, A. M. (2007). Development and Evaluation of a Core Subset of the USDA Rice Germplasm Collection. Crop Science, *47*(2), 869-876. doi:https://doi.org/10.2135/cropsci2006.07.0444

[66] Anderson, J. V., & Morris, C. F. (2001). An Improved Whole-Seed Assay for Screening Wheat Germplasm for Polyphenol Oxidase Activity. Crop Science, *41*(6), 1697-1705. doi:https://doi.org/10.2135/cropsci2001.1697

[67] Upadhyaya, H. D., Bramel, P. J., Ortiz, R., & Singh, S. (2002). Developing a Mini Core of Peanut for Utilization of Genetic Resources. Crop Science, *42*(6), 2150-2156. doi:https://doi.org/10.2135/cropsci2002.2150

[68] Langpap, C., & Wu, J. (2004). Voluntary conservation of endangered species: when does no regulatory assurance mean no conservation? Journal of Environmental Economics and Management, *47*(3), 435-457.

[69] Hinze, L. L., Dever, J. K., & Percy, R. G. (2012). Molecular Variation Among and Within Improved Cultivars in the U.S. Cotton Germplasm Collection. *Crop Science, 52*(1), 222-230. doi:https://doi.org/10.2135/cropsci2011.04.0202

[70] Drechsler, M., Wätzold, F., Johst, K., Bergmann, H., & Settele, J. (2007). A model-based approach for designing cost-effective compensation payments for conservation of endangered species in real landscapes. Biological conservation, *140*(1-2), 174-186.

Chapter 8

Climate Change Vulnerability Assessment of Imperiled Plants in the Mojave Desert

Jennifer Wilkening, Lara Kobelt and Tiffany J. Pereira

Abstract

The Mojave Desert in the southwestern United States is one of the hottest and driest areas of North America. Climate change is likely to exacerbate these conditions. The region is home to many endemic plant species, including 24 federally threatened species. The impact of climate change factors on these sensitive Mojave Desert species is relatively unknown. Here we used a climate change vulnerability assessment to determine which imperiled plants may be most affected by changing climatic conditions. We evaluated the vulnerability of each species under future climate scenarios and calculated scores using metrics such as exposure, sensitivity, niche breadth, and dispersal capability. We found that most listed plant species were vulnerable to climate change, with 21% (N = 5) classified as extremely vulnerable, 25% (N = 6) classified as highly vulnerable, and 42% (N = 10) classified as moderately vulnerable. Contributing factors most frequently associated with vulnerability included various barriers to migration, high habitat specificity, and species sensitivity to changes in hydrological patterns. Many of these species are already threatened by ongoing anthropogenic stressors such as urban growth and associated developments, and these results suggest that climate change will pose additional challenges for conservation and management. Natural resource managers can use the vulnerability ranking and contributing factors identified from these analyses to inform ecological decisions related to threatened plants throughout desert regions.

Keywords: threatened species, climate change, Mojave Desert, vulnerability assessment, conservation management, endemic species, biodiversity

1. Introduction

The Mojave Desert occurs across parts of California, Nevada, Utah, and Arizona in the western region of the United States. It is the smallest and driest desert in North America, with annual precipitation varying widely across the region and an average maximum temperature of 36°C in the summer and average minimum temperature of 13°C in the winter [1]. Although it is not the planet's hottest desert, it does hold the record for the highest land surface temperature ever recorded on Earth (56.7°C), registered at Death Valley National Park in 1913 [2]. The region is ecologically and topographically diverse, with elevations ranging from to −85 to 3,633 meters, and consequently hosts a large number of threatened, endangered, and endemic species. The sunny climate and proximity to national parks and other public lands also attracts human inhabitants. Most of the large cities in the area (e.g., Las Vegas) continue to grow, and anthropogenic developments and recreational activities are increasingly disturbing Mojave Desert ecosystems. Additional stressors on native flora and fauna include invasive species, altered disease dynamics, and increasing wildfire frequency, and populations of many plant and animal species are in decline.

As a patchwork of unique habitats and extreme conditions, the Mojave Desert supports a diverse flora of about three thousand plant species [3]. The number and geological distribution of plants within this landscape is determined by the interaction of many factors such as elevation, precipitation, temperature, soil properties and disturbance. Many Mojave plant species are both regionally endemic, found nowhere else on earth, and locally endemic, restricted to a specific geographic area within the Mojave Desert. Additionally, some species can be defined as rare, having either a narrow geographical range, specialized habitat requirements, or a small population size [4]. Although a species may be considered endemic but not rare, and vice versa, both endemic and/or rare species exhibit attributes which make them more vulnerable to extinction [5]. Consequently, conservation efforts are necessary in some instances to preserve genetic and species diversity as well as ecosystem structure and function [3]. Given the number of endemic and rare species, and their unique habitat requirements, land managers face a daunting task in the Mojave Desert. The United States Endangered Species Act (ESA) of 1973 provided a critical boost to conservation efforts. This legislation was designed to identify and protect plant and animal species facing imminent extinction, and covers species, subspecies, and distinct population segments [6]. The Mojave Desert contains 24 plant species listed as threatened or endangered under the ESA (**Table 1**), many of which are also listed under some level of conservation concern by the International Union for the Conservation of Nature [7].

These 24 listed plants species encompass four states and nine plants families, illustrating the starkly different geographical reaches of the Mojave Desert. They can be further subdivided to exemplify one or more of the three criteria for rarity [4] and/or specific forms of endemism [8]. First, all the listed species have a narrow geographic range, some restricted to soils derived from specific geological formations and some endemic to regional biodiversity hotspots. For example, the seven species from Nevada are all found within the confines of Ash Meadows National Wildlife Refuge and land immediately adjacent (**Figure 1**). This 24,000-acre spring-fed alkaline wetland and alkaline desert upland has the highest concentration of endemic species in the United States [9–10]. Second, all the listed species have highly specialized habitat requirements ranging from eolian dunes of Coachella Valley, California to gypsum outcrops associated with the Harrisburg Member of the Kaibab Formation in northern Mohave County, Arizona. Edaphic endemism (influenced by soils; [8]) is a strong theme running through the Mojave ESA listed species with alkaline-wetland, calcium-carbonate, gypsum, and granite endemism represented. Three listed species from the San Bernardino Mountains in California are restricted to "pebble plains" named for the layer of orange quartzite pebbles that are pushed to the clay soil surface by freezing and thawing [11]. Third, many of the listed species have small or disjointed populations, such as Lane Mountain Milkvetch (*Astragalus jaegerianus*). This species is known from only 4 disjunct populations and is threatened by recreational off-highway vehicle (OHV) use and military activity. Although threats to the 24 ESA listed Mojave Desert plant species are many, climate change remains a ubiquitous concern.

Scientific Name	Common Name	Family	US ESA Status	Nature Serve Status	State(s) of occurrence
Arctomecon humilis Coville	Dwarf Bear-poppy	Papaveraceae	E	G1	Utah
Arenaria ursina B.L. Rob.	Bear Valley Sandwort	Caryophyllaceae	T	G1	California
Astragalus albens Greene	Cushenbury Milkvetch	Fabaceae	E	G1	California
Astragalus ampullarioides Sheldon	Shivwits Milkvetch	Fabaceae	E	G1	Utah
Astragalus holmgreniorum Barneby	Holmgren's Milkvetch	Fabaceae	E	G1	Arizona, Utah
Astragalus jaegerianus Munz	Lane Mountain Milkvetch	Fabaceae	E	G2	California
Astragalus lentiginosus Douglas var. *coachellae* Barneby	Coachella Valley Milkvetch	Fabaceae	E	T1	California
Astragalus phoenix Barneby	Ash Meadows Milkvetch	Fabaceae	T	G2	Nevada
Astragalus tricarinatus A. Gray	Triple-rib Milkvetch	Fabaceae	E	G2	California
Castilleja cinerea A. Gray	Ash Gray Indian-paintbrush	Scrophulariaceae	T	G1	California
Centaurium namophilum Reveal, C.R. Broome & Beatley	Spring-loving Centaury	Gentianaceae	T	G2	Nevada
Enceliopsis nudicaulis (A. Gray) A. Nelson var. *corrugata* Cronquist	Ash Meadows Sunray	Asteraceae	T	T1	Nevada
Erigeron parishii A. Gray	Parish's Daisy	Asteraceae	T	G2	California
Eriogonum kennedyi Porter ex *S. Watson* var. *austromontanum* Munz & I. M. Johnst.	Southern Mountain Buckwheat	Polygonaceae	T	T2	California
Eriogonum ovalifolium Nutt. var. *vineum* (Small) A. Nelson	Cushenbury Buckwheat	Polygonaceae	E	T1	California
Grindelia fraxinipratensis Reveal & Beatley	Ash Meadows Gumweed	Asteraceae	T	G2	Nevada
Ivesia kingii S. Watson var. *eremica* (Coville) Ertter	Ash Meadows Ivesia	Rosaceae	T	T1	Nevada
Lesquerella kingii S. Watson ssp. *bernardina* (Munz) Munz	San Bernardino Mountains Bladderpod	Brassicaceae	E	T1	California

Scientific Name	Common Name	Family	US ESA Status	Nature Serve Status	State(s) of occurrence
Mentzelia leucophylla Brandegee	Ash Meadows Blazingstar	Loasaceae	T	G1	Nevada
Nitrophila mohavensis Munz & J. C. Roos	Amargosa Niterwort	Chenopodiaceae	E	G1	California, Nevada
Oxytheca parishii Parry var. *goodmaniana* Ertter	Cushenbury Oxytheca	Polygonaceae	E	T1	California
Pediocactus sileri (Engelm. ex J.M. Coult.) L.D. Benson	Siler Pincushion Cactus	Cactaceae	T	G2	Arizona, Utah
Sphaeralcea gierischii N.D. Atwood & S.L. Welsh	Gierisch's Globemallow	Malvaceae	E	G1	Arizona, Utah
Swallenia alexandrae (Swallen) Söderst. & Decker	Eureka Dune Grass	Poaceae	E	G2	California

Global conservation status rank according to NatureServe: GX/TX = presumed extinct, GH/TH = possibly extinct, G1/T1 = critically imperiled, G2/T2 = imperiled, G3/T3 = vulnerable, G4/T4 = apparently secure, G5/T5 = secure.

Table 1.
Name, family, and location of Mojave Desert plants listed as threatened (T) or endangered (E) under the United States Endangered Species Act (ESA).

There has been multi-decadal warming across the Earth's surface, with each of the previous three decades experiencing progressively warmer temperatures than any preceding decade since 1850 [12]. Ecoregions defined by climatic extremes, such as deserts, are particularly vulnerable to climate change. On a global scale, deserts have experienced faster warming and drying over the last 50 years when compared to other regions [13, 14], and this is projected to continue [15]. Desert dwelling species may be directly impacted by thermal or hydric stress, or indirectly via altered habitats, species interactions, or disease dynamics. Several studies have documented negative effects from increasing temperatures on desert wildlife species, including birds [16, 17], mammals [18], invertebrates [19], and reptiles [20]. Studies focused on desert plants have predicted species range shifts in response to climate change [21], and potential changes in vegetation community composition as a result of altered summer precipitation patterns [22]. Climate change impacts are projected to be particularly severe in the southwestern region of the United States, where the Mojave Desert occurs, and resident plants will have to cope with prolonged drought, fewer frost days, warmer temperatures, and an increase in extreme weather events [23–25]. Many species may not be able to survive in such highly transformed environments. Climate change has emerged as one of the greatest threats to biodiversity, with potential to hasten species extinctions, elevating the need to understand how threatened and endangered species may be affected [26]. Management actions have traditionally focused on the establishment of protected areas to conserve habitat and halt species decline, but the efficacy of these strategies is questionable under predicted future climatic regimes [27]. One way to better integrate climate change considerations into management planning for imperiled plant species is to conduct climate change vulnerability assessments (CCVAs).

Vulnerability assessments offer a standardized approach for measuring climate change sensitivity that is efficient, repeatable, and directly comparable among

Figure 1.
Illustration panel by Tiffany J. Pereira portraying the federally threatened and endangered plants of Ash Meadows National Wildlife Refuge, a vulnerable biodiversity hotspot in Nevada, USA. From left to right: Centaurium namophilum, Astragalus phoenix, Enceliopsis nudicaulis var. corrugata, Grindelia fraxinipratensis, Ivesia kingii var. eremica, *and* Mentzelia leucophylla. *Not pictured:* Nitrophila mohavensis.

different species. Vulnerability can be defined as the degree to which a species is susceptible to climate change, taking into consideration the magnitude of predicted change where the species occurs and the adaptive capacity of the species [28]. Species within an area or ecoregion can be ranked according to their expected sensitivity to changing climatic conditions [29]. A CCVA can be conducted using several different methods, such as a correlative, mechanistic, or trait-based approach, or a combination of several of these types of models. Correlative approaches have frequently been used for plants, but a trait-based approach is best for rapid assessment of a larger number of species [30]. A trait-based method identifies and scores attributes of a species that are relevant to avoiding or tolerating climate change, such as dispersal ability or sensitivity to temperature or precipitation changes. A species' overall vulnerability to climate change is represented by the sum of these scores. Here we utilize a trait-based approach to assess climate change vulnerability of threatened and endangered plants in the Mojave Desert, and provide suggestions for incorporating these considerations into management actions and conservation planning efforts.

2. Methods

We accessed the United States Fish and Wildlife Service's (USFWS) website to obtain the most up to date list of plant species identified as threatened or endangered under the ESA [31]. This list was reduced to only those species occurring in the Mojave Desert, as defined by the EPA Level III Ecoregions [32]. We overlaid spatial data of plant species distribution with a boundary of the selected ecoregion? (Mojave Basin and Range) to verify species occurrence, which resulted in 24 plant species in the Mojave Desert listed as threatened or endangered (**Figure 2**).

We used NatureServe's Climate Change Vulnerability Index (https://www.natureserve.org/conservation-tools/climate-change-vulnerability-index, last accessed 19 Nov 2020) to investigate the susceptibility of Mojave Desert listed plants to climate change (CCVI) [33]. This index incorporated future climate projections, along with data for each species related to natural history, current distribution, and ecological associations, to predict range contraction and/or population extirpations. The CCVI used a total of 24 factors in several categories

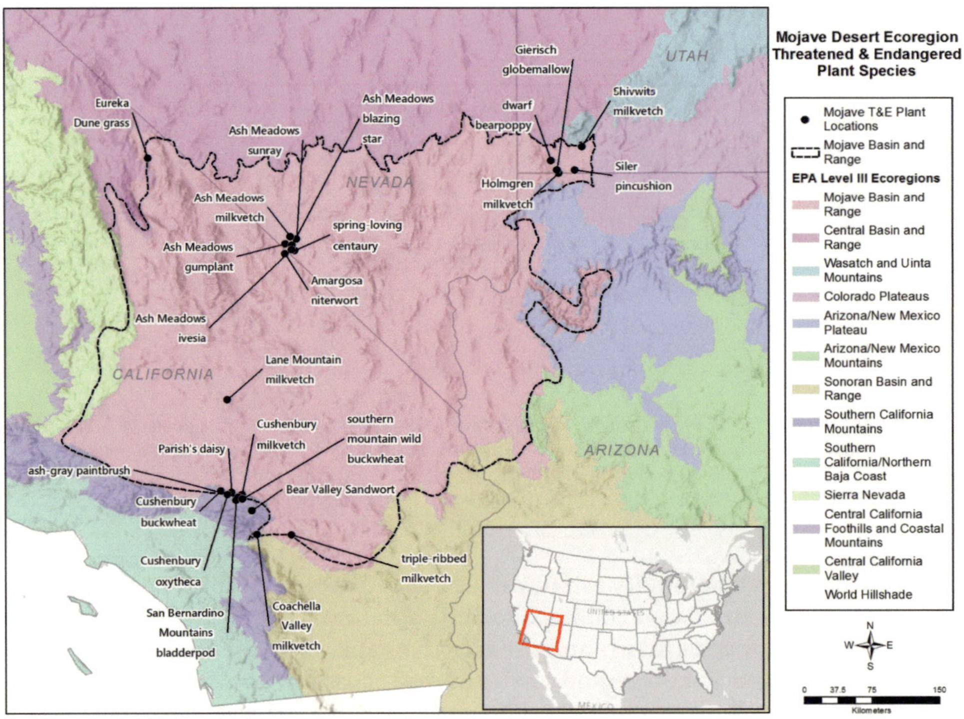

Figure 2.
Plant species listed as threatened or endangered under the United States Endangered Species Act in the Mojave Desert.

Extremely Vulnerable (EV): Abundance and/or range extent within geographical area assessed extremely likely to substantially decrease or disappear by 2050.
Highly Vulnerable (HV): Abundance and/or range extent within geographical area assessed likely to decrease significantly by 2050.
Moderately Vulnerable (MV): Abundance and/or range extent within geographical area assessed likely to decrease by 2050.
Less Vulnerable (LV): Available evidence does not suggest that abundance and/or range extent within the geographical area assessed will change (increase/decrease) substantially by 2050. Actual range boundaries may change.
Insufficient Evidence (IE): Information about a species' vulnerability is inadequate to calculate an index score.

Table 2.
Climate Change Vulnerability Index Score Descriptions [33].

to assess the three major elements of vulnerability; exposure to climate change (direct and indirect), sensitivity, and species-specific adaptive capacity. We reviewed information about the species and entered a score for each factor according to guidance and criteria set by the CCVI [33]. Direct climate exposure was calculated by estimating projected temperature and moisture change within the assessment area, while indirect exposure was assessed by evaluating future distribution relative to barriers that may restrict a species ability to shift its' range. Mid-century (2040–2069) climate projections were used based on an ensemble-average of general circulation models and a medium emissions scenario [33]. For sensitivity and adaptive capacity, we evaluated a variety of factors including dispersal capability, reliance on specific thermal or hydrological conditions, dependence on disturbance regime or snow/ice cover, restriction to uncommon habitat types or landscape features, reliance on interspecific interactions, and genetic variation. For each sensitivity factor, we assigned a score of decrease, somewhat decrease, neutral, somewhat increase, increase, or greatly increase vulnerability to climate change. A score of unknown was given when information was lacking for a particular factor. Although some sensitivity factors were optional, the CCVI required a minimum number of factors (10) in order to avoid a determination of insufficient evidence [33].

Each species was assigned one of the following five categories based on CCVI scores: extremely vulnerable, highly vulnerable, moderately vulnerable, less vulnerable, and insufficient evidence (**Table 2**). The CCVI also calculated a categorical confidence estimate (very low, low, high, very high) for each species ranking, which was based on certainty in the factor values as represented by the frequency of multiple categories of vulnerability being selected for a given factor.

3. Results

The majority of the species assessed (N = 21, 88%) were found to be moderately, highly, or extremely vulnerable to climate change (**Figure 3**). Five species (21%) were extremely vulnerable to climate change, six species (25%) were highly vulnerable, and ten species (42%) were moderately vulnerable to climate change (**Figure 3**). Only three species were determined to be less vulnerable to climate change and no species received a score of insufficient evidence. The confidence

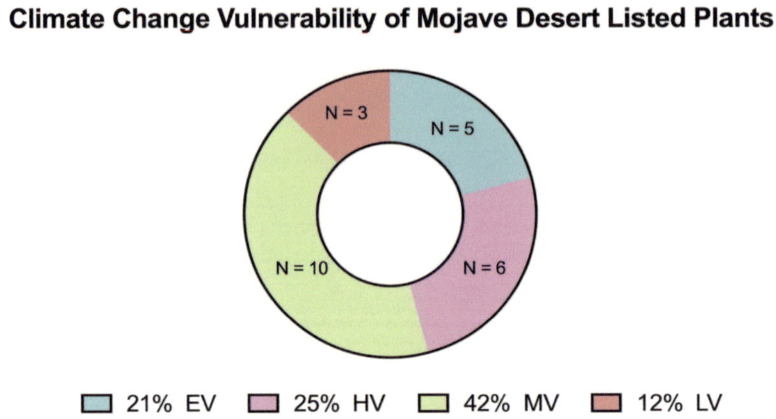

Figure 3.
Proportion of Mojave Desert listed plants categorized by the following climate change vulnerability indices: EV = extremely vulnerable, HV = highly vulnerable, MV = moderately vulnerable, LS = less vulnerable. N is the number of species within each category, out of 24 species total.

estimate for each species score was very high, indicating a large degree of certainty in the vulnerability ranking.

Among key factors, limited dispersal capability increased climate change vulnerability for almost every species assessed (N = 23), with the one exception being the only grass species, Eureka Dune Grass (*Swallenia alexandrae*), which is wind pollinated (**Table 3**). Likewise, natural or anthropogenic barriers that may impede range shifts, increased or somewhat increased climate change vulnerability for the majority of plant species assessed (N = 22, **Table 3**). Physical habitat, or restriction to uncommon geological formations or substrates, also emerged as one of the most important factors in our assessment, as it somewhat increased or increased vulnerability for most species (N = 22). The potential for climate change mitigation projects (e.g., alternative energy facilities) also somewhat increased vulnerability for many species (N = 15, **Table 3**), which is not surprising given that these desert plants inhabit areas often identified as prime locations for solar energy projects. Additionally, physiological hydrological niche was a key factor affecting vulnerability ranking in many species (N = 9), with species reliant on a particular hydrologic regime (e.g., desert springs) being assessed as more vulnerable than species not dependent on these habitats. Competition from other native or non-native species favored by climate change somewhat increased vulnerability for roughly half of the species assessed (N = 11). Genetic variation appeared to be a less important factor (N = 1), similar to reproductive system which somewhat increased vulnerability in only a few species (N = 6, **Table 3**).

The five species classified as extremely vulnerable occurred in the same geographic area (southwest Utah and northwest Arizona, **Figure 2**). Among species classified as highly vulnerable (N = 6), five of these were located primarily in a protected area at Ash Meadows National Wildlife Refuge in Nevada (**Table 3**). Species classified as less vulnerable (N = 3) also included one that inhabited a protected area (**Table 3**). Roughly 38% of the species assessed (N = 9) were located exclusively or primarily on lands under protection status, with seven species endemic to Ash Meadows National Wildlife Refuge in Nevada, one species occurring in Death Valley National Park in California, and one species in Zion National Park in Utah. In addition to climate change, many plant species were also found to be threatened by other anthropogenic factors such as agriculture, grazing, groundwater pumping, invasive species, mining, recreational off-highway vehicle (OHV) use and activities, and urban development (**Table 3**).

Common name	CCVI	Key factors	Occurs in Protected Area	Additional Anthropogenic Stressors
Dwarf Bear-poppy	EV	Natural barriers, climate change mitigation, dispersal, physical habitat, pollinators, reproductive system	No	mining, power line construction, recreational OHV use, road construction, urban development
Shivwits Milkvetch	EV	Anthropogenic barriers, climate change mitigation, dispersal, physical habitat, competition, reproductive system	Yes	agriculture, grazing, urban development
Holmgren's Milkvetch	EV	Natural barriers, climate change mitigation, dispersal, physical habitat, competition, reproductive system	No	grazing, mineral development, power line construction, recreational OHV use and activities, road construction, urban development
Siler Pincushion Cactus	EV	Natural barriers, anthropogenic barriers, climate change mitigation, dispersal, physical habitat	No	grazing, mining, oil pipeline construction, recreational OHV use, urban development
Gierisch's Globemallow	EV	Natural barriers, climate change mitigation, dispersal, physical habitat, competition	No	grazing, invasive species, mining, recreational OHV use
Bear Valley Sandwort	HV	Natural barriers, anthropogenic barriers, dispersal, physiological hydrological niche, physical habitat, competition, reproductive system	No	mining, grazing, recreational OHV use and activities, urban development
Ash Meadows Milkvetch	HV	Natural barriers, climate change mitigation, dispersal, physiological hydrological niche, physical habitat, pollinators	Yes	agriculture, grazing, urban development
Ash Meadows Gumweed	HV	Natural barriers, climate change mitigation, dispersal, physiological hydrological niche, physical habitat, competition	Yes	agriculture, grazing, groundwater pumping, invasive species, mining, recreational OHV use, road construction

Common name	CCVI	Key factors	Occurs in Protected Area	Additional Anthropogenic Stressors
Ash Meadows Ivesia	HV	Natural barriers, climate change mitigation, dispersal, physiological hydrological niche, physical habitat, competition	Yes	agriculture, grazing, groundwater pumping, mining
Ash Meadows Blazingstar	HV	Natural barriers, climate change mitigation, dispersal, physiological hydrological niche, physical habitat	Yes	agriculture, groundwater pumping, mining
Amargosa Niterwort	HV	Natural barriers, climate change mitigation, dispersal, physiological hydrological niche, physical habitat	Yes	agriculture, grazing, groundwater pumping, mining, recreational OHV use, road construction
Cushenbury Milkvetch	MV	Climate change mitigation, dispersal, physical habitat	No	mining, recreational OHV use and activities, road construction, urban development
Lane Mountain Milkvetch	MV	Climate change mitigation, dispersal, competition, genetic variation	No	military activities, mineral collecting, mining, recreational OHV use
Coachella Valley Milkvetch	MV	Anthropogenic barriers, climate change mitigation, dispersal, physical habitat, competition	No	flood control projects, recreational OHV use, road construction, urban development
Ash Gray Indian-paintbrush	MV	Natural barriers, dispersal, physiological hydrological niche, physical habitat, competition, reproductive system	No	grazing, invasive species, mining, recreational development (e.g., ski resorts), recreational OHV use, urban development
Spring-loving Centaury	MV	Natural barriers, climate change mitigation, dispersal, physiological hydrological niche, physical habitat	Yes	agriculture, groundwater pumping, mining, recreational OHV use, urban development

Common name	CCVI	Key factors	Occurs in Protected Area	Additional Anthropogenic Stressors
Ash Meadows Sunray	MV	Natural barriers, climate change mitigation, dispersal, physiological hydrological niche, physical habitat	Yes	agriculture, groundwater pumping, mining, recreational OHV use, road construction
Parish's Daisy	MV	Natural barriers, dispersal, physical habitat	No	grazing, mining, recreational OHV use, urban development
Southern Mountain Buckwheat	MV	Natural barriers, dispersal, physical habitat, reproductive system	No	invasive species, mining, recreational development (e.g., ski resorts), recreational OHV use, urban development
San Bernardino Mountains Bladderpod	MV	Natural barriers, dispersal, physical habitat, competition	No	grazing, invasive species, mining, recreational development (e.g., ski resorts), recreational OHV use, urban development
Cushenbury Oxytheca	MV	Natural barriers, dispersal, physical habitat	No	hydroelectric development, mining, power line construction, recreational development (e.g., ski resorts), recreational OHV use, urban development
Triple-rib Milkvetch	LV	Natural barriers, dispersal, competition	No	grazing, oil pipeline construction, recreational OHV use
Cushenbury Buckwheat	LV	Natural barriers, anthropogenic barriers, dispersal, physical habitat	No	hydroelectric development, mining, power line construction, recreational development (e.g., ski resorts), recreational OHV use, urban development

Common name	CCVI	Key factors	Occurs in Protected Area	Additional Anthropogenic Stressors
Eureka Dune Grass	LV	Natural barriers, physical habitat	Yes	recreational OHV use and activities

Key factors contributing to vulnerability for each species. Factors related to indirect climate exposure include natural and anthropogenic barriers that restrict species movement in response to changing climatic conditions, and the degree to which the species may be affected by climate change mitigation actions (ie, placement of solar arrays in desert plant habitat). All other listed key factors relate to species sensitivity and adaptive capacity. Dispersal refers to the ability to move through unsuitable habitat, while physiological hydrological niche pertains to plant species dependent upon a narrowly defined water source (e.g., desert springs). Physical habitat refers to a species dependence upon a particular uncommon landscape or geological feature (e.g., gypsiferous soils). Pollinators relates to pollination strategy (e.g., dependence on only one pollinator may increase vulnerability), while competition identifies species that may be outcompeted by another species (native or non-native) favored by climate change. Reproductive system serves as a measure of genetic diversity in plants where genetic information is lacking, and genetic variation serves as a proxy for a species capacity to adapt to novel conditions. Protection area status is indicated by whether or not a species occurs exclusively or primarily in a protected area (e.g., wildlife refuge, park, area of environmental concern). Non-climate related anthropogenic stressors that are negatively affecting plant populations are also displayed.

Table 3.
Climate change vulnerability index score for threatened and endangered plants in the Mojave Desert (EV = extremely vulnerable, HV=highly vulnerable, MV = moderately vulnerable, LV = less vulnerable).

4. Discussion

Our results suggest that climate change may pose additional threats, or exacerbate current stressors, for threatened and endangered plants in the Mojave Desert. Many listed plants are rare species characterized by relatively narrow ecological niches, small population sizes and restricted geographic ranges [34]. Although these traits may contribute to a species decline in a world increasingly influenced by human activity, these factors alone do not determine likelihood of extinction. Numerous species that were once locally common or abundant (e.g., elephants, lemurs, orangutans) have become endangered due to hunting, habitat loss, agriculture, or other human endeavors. Rarity may have been a contributing factor, but Mojave Desert listed plants became threatened or endangered primarily because of anthropogenic stressors. Climate change has not been considered in measures of extinction risk in the past, and this phenomenon has the potential to accelerate extinction processes for species already struggling to persist. For example, the five plant species classified as extremely vulnerable by our CCVI (Dwarf Bear-poppy, Shivwits Milkvetch, Holmgren's Milkvetch, Siler Pincushion Cactus, Gierisch's Globemallow) occur only in a small geographic area located in southern Utah and northern Arizona. Climate change is projected to be particularly severe for this part of the Mojave Desert, with a 2.2°C–4.4°C increase in annual temperature predicted by the end of the century (2070–2099) and a 30% reduction in snowpack, which functions as the regions' main source of water [35].

Although Shivwits Milkvetch occurs predominantly in a protected area (Zion National Park), the remaining four plants categorized as extremely vulnerable do not. These species are habitat specialists restricted to gypsiferous soils located largely on federally managed public lands where populations are also threatened by recreation, grazing, and gypsum mining. One frequent suggestion for managing imperiled species under climate change is to reduce existing threats in order to increase resilience to climate change [36]. Threats could be minimized by placing core habitat areas under protection status, which has been successful in recovery efforts for another listed plant in the Mojave Desert, the Eureka Dune Grass. The species was recently down listed from endangered to threatened, mainly as a result of habitat areas becoming part of Death Valley National Park which prohibits OHV use in Eureka Dunes [37]. This strategy may not be effective throughout the Mojave Desert, however, as listed plant species in Ash Meadows National Wildlife Refuge continue to decline despite similar protection status [38]. Endemic species in this refuge rely upon rare, wet microhabitats sustained by desert springs or shallow groundwater, which are vulnerable to groundwater pumping and drought. Groundwater pumping is very likely to increase throughout the Mojave Desert as droughts become more frequent and dry conditions are intensified by climate change, and the future of many Ash Meadows species remains uncertain.

Similar to other studies related to CCVAs of imperiled plants, our vulnerability scores were strongly associated with two analogous factors, natural or anthropogenic barriers and dispersal ability. In highly urbanized and densely populated regions of California, anthropogenic barriers limited the dispersal capability of 63% of plant species assessed [39]. A lack of topographical variation presented a bigger challenge in other regions, as plant species inhabiting areas of topographical homogeneity may experience climate change effects sooner since they are unable to move up or down in elevation [40]. For plants in relatively flat regions of the United States, such as Illinois, latitudinal migration may be the only effective survival mechanism, which will require assisted migration as individuals will not be able to disperse through highly urbanized areas [40]. In contrast, the Mojave Desert is relatively undeveloped and topographically diverse, thus offers opportunities for plants to shift to higher elevations and more mesic microhabitats. Hot, dry desert valleys and other areas of unsuitable habitat may present similar barriers to dispersal, however. *Some* of the plant species assessed here *could* be good candidates for assisted migration, assuming appropriate substrates and ecological conditions (e.g., pollinators, hydrological regimes) are available on the landscape and successful restoration techniques are established.

5. Conclusion

Information produced by the CCVA can be used to identify threatened and endangered plants most vulnerable to climate change in the Mojave Desert. Species vulnerability assessments break down the complexity of climate change impacts on overall biodiversity, and facilitate the integration of societal, economic, and other environmental concerns into conservation planning efforts. Our assessments also identified knowledge gaps for each species, which promotes the development and testing of new hypotheses about climatic tolerances. Furthermore, the identification of particular traits that make a species vulnerable allow for targeted management actions. Recovery and management plans for threatened or endangered species can be updated to include climate change vulnerability and its implications, which may necessitate the inclusion of different stakeholders or increased frequency of monitoring to detect distribution shifts. The Mojave Desert is a unique ecoregion beloved by many and there is still much uncertainty related to the magnitude and extent of global atmospheric and climatic change. Our hope is that results presented here contribute to the larger body of knowledge for the region, and aid in better stewardship of these irreplaceable ecological systems and inhabitants.

Acknowledgements

The findings and conclusions in this article are those of the author(s) and do not necessarily represent the views of the U.S. Fish and Wildlife Service or the U.S. Bureau of Land Management. We thank Gary Reese and Cayenne Engel from the Nevada Division of Forestry for their conscientious review and helpful comments which greatly improved the manuscript.

Author details

Jennifer Wilkening[1*], Lara Kobelt[2] and Tiffany J. Pereira[3]

1 United States Fish and Wildlife Service, Las Vegas, Nevada, USA

2 United States Bureau of Land Management, Las Vegas, Nevada, USA

3 Desert Research Institute, Las Vegas, Nevada, USA

*Address all correspondence to: jennifer_wilkening@fws.gov

References

[1] US Climate Summaries Mojave Desert [Internet]. 2020. Available from: https://www.usclimatedata.com/ [Accessed December 2020]

[2] World Meteorological Organization. World: Highest Temperature [Internet]. 2020. Available from: https://wmo.asu.edu/content/world-highest-temperature [Accessed October 2020]

[3] Walker LR, Landau FH. 2018. A natural history of the Mojave Desert. University of Arizona Press.

[4] Primack RB. 2002. Essentials of Conservation Biology. Sinauer Assoc., Inc. Sunderland, MA.

[5] Sodhi NS, Brook BW, Bradshaw, CJ. 2009. Causes and consequences of species extinctions. The Princeton Guide to Ecology 1:514-520.

[6] USFWS (U.S. Fish and Wildlife Service) and NMFS (National Marine Fisheries Service). 1996. Policy regarding the recognition of distinct vertebrate population segments under the Endangered Species Act. Federal Register 61:4721-4725.

[7] International Union for Conservation of Nature [Internet]. 2020. The IUCN Red List of Threatened Species. Version 2020-2. Available from: http://www.iucnredlist.org [Accessed October 2020]

[8] Mason HL. 1946. The edaphic factor in narrow endemism. II. the geographic occurrence of plants of highly restricted patterns of distribution. Madrono 8:241-257.

[9] Pister EP. 1974. Desert fishes and their habitats: Transactions of the American Fisheries Society 103:531-540.

[10] Sada DW. 1990. Recovery plan for the endangered and threatened species of Ash Meadows, Nevada: Reno, Nevada. U.S. Fish and Wildlife Service.

[11] USFWS (U.S. Fish and Wildlife Service). 2006. Designation of Critical Habitat for *Arenaria ursina* (Bear Valley sandwort), *Castilleja cinerea* (ash-gray Indian paintbrush), and *Eriogonum kennedyi var. austromontanum* (southern mountain wild-buckwheat); Proposed Rules. Federal Register 71: 67712-67754.

[12] International Panel on Climate Change. 2014. Climate Change 2014: Impacts, Adaptation, and Vulnerability. Summary for Policymakers. Cambridge University Press, Cambridge, United Kingdom and New York, NY, USA. 44 pp. URL: http://ipcc wg2.gov/AR5/images/uploads/IPCC_WG2AR5_SPM_Approved.pdf

[13] Zhou L, Chen H, Dai Y. 2015. Stronger warming amplification over drier ecoregions observed since 1979. Environmental Research Letters 10:064012

[14] Wuebbles, DJ, Fahey DW, Hibbard KA. 2017. Climate science special report: fourth national climate assessment, volume I.

[15] Dominguez F, Cañon J, Valdes J. 2010. IPCC-AR4 climate simulations for the Southwestern US: the importance of future ENSO projections. Climatic Change 99:499-514.

[16] du Plessis KL, Martin RO, Hockey PA, Cunningham SJ, Ridley AR. 2012. The costs of keeping cool in a warming world: implications of high temperatures for foraging, thermoregulation and body condition of an arid-zone bird. Global Change Biology 18:3063-3070.

[17] Iknayan KJ, Beissinger SR. 2018. Collapse of a desert bird community over the past century driven by climate change. Proceedings of the National Academy of Sciences 115:8597-8602.

[18] Moses MR, Frey JK, Roemer GW. 2012. Elevated surface temperature depresses survival of banner-tailed kangaroo rats: will climate change cook a desert icon? Oecologia 168:257-268.

[19] Crawford CS (1981) Biology of Desert Invertebrates (Springer, New York).

[20] Sinervo B, Mendez-De-La-Cruz F, Miles DB, Heulin B, Bastiaans E, Villagrán-Santa Cruz M, Lara-Resendiz R, Martínez-Méndez N, Calderón-EspinosaML,Meza-LázaroRN, Gadsden H. 2010. Erosion of lizard diversity by climate change and altered thermal niches. Science 328:894-899.

[21] Kelly AE, Goulden ML. 2008. Rapid shifts in plant distribution with recent climate change. Proceedings of the National Academy of Sciences 105:11823-11826.

[22] Ehleringer JR, Phillips SL, Schuster WS, Sandquist DR. 1991. Differential utilization of summer rains by desert plants. Oecologia 88:430-434.

[23] Archer SR, Predick KI. 2008. Climate change and ecosystems of the southwestern United States. Rangelands 30:23-28.

[24] Garfin G, Franco G, Blanco H, Comrie A, Gonzalez P, Piechota T et al. 2014. Southwest. In: Melillo JM, Richmond TC, Yohe GW editors. Climate Change Impacts in the United States: The Third National Climate Assessment. United States Global Change Research Program. P. 462-486.

[25] Cook BI, Ault TR, Smerdon JE. 2015. Unprecedented 21st century drought risk in the American Southwest and Central Plains. Science Advances 1: e1400082.

[26] Dirzo R, Young HS, Galetti M, Ceballos G, Isaac NJ, Collen B. 2014. Defaunation in the Anthropocene. Science 345:401-406.

[27] Langdon JGR, Lawler JJ. 2015. Assessing the impacts of projected climate change on biodiversity in the protected areas of western North America. Ecosphere 6:1-14.

[28] Glick P, Stein BA, Edelson NA. 2011. Scanning the conservation horizon: A guide to climate change vulnerability assessment. Washington, DC: National Wildlife Federation.

[29] Gardali T, Seavy NE, DiGaudio RT, Comrack LA. 2012. A climate change vulnerability assessment of California's at-risk birds. PLoS One 7 p.e29507.

[30] Foden WB, Young BE, Akçakaya HR, Garcia RA, Hoffmann AA, Stein BA, Thomas CD, Wheatley CJ, Bickford D, Carr JA, Hole DG. 2019. Climate change vulnerability assessment of species. Wiley Interdisciplinary Reviews: Climate Change 10 p.e551.

[31] United States Fish and Wildlife Service. FWS-Listed U.S. Species by Taxonomic Group - All Plants [Internet]. 2020. Available from: https://ecos.fws.gov/ecp/report/species-listings-by-tax-group?status Category=Listed&groupName=All%20 Plants [Accessed July 2020]

[32] United States Environmental Protection Agency. Level III and IV Ecoregions of the Continental United States [Internet]. 2020. Available from: https://www.epa.gov/eco-research/level-iii-and-iv-ecoregions-continental-united-states [Accessed November 2020]

[33] Young BE, Byers E, Hammerson G, Frances A, Oliver L, Treher A. 2016. Guidelines for using the NatureServe Climate Change Vulnerability Index. NatureServe, Arlington, Virginia, USA.

[34] Drury WH. 1974. Rare species. Biological Conservation 6:162-169.

[35] Khatri KB, Strong C. Climate Change, Water Resources, and Potential Adaptation Strategies in Utah. Department of Water Resources, Utah Department of Natural Resources. Salt Lake City, Utah, USA.

[36] Mantyka-pringle CS, Martin TG, Rhodes JR. 2012. Interactions between climate and habitat loss effects on biodiversity: a systematic review and meta-analysis. Global Change Biology 18:1239-52.

[37] USFWS (U.S. Fish and Wildlife Service). 2018. Endangered and Threatened Wildlife and Plants; Removing *Oenothera avita ssp. eurekensis* From the Federal List of Endangered and Threatened Plants, and Reclassification of *Swallenia alexandrae* From Endangered to Threatened. Federal Register 83:8576-8603.

[38] Moore-O'Leary K, Levine C. 2019. Protocol Survey Report 2014-2016: Monitoring of Nine Endemic Rare Plants (FF08RASH00-053). United States Fish and Wildlife Service, Region 8 Inventory and Monitoring Program. 63 pp.

[39] Anacker BL, Gogol-Prokurat M, Leidholm K, Schoenig S. 2013. Climate change vulnerability assessment of rare plants in California. Madroño 60:193-210.

[40] Molano-Flores B, Zaya DN, Baty J, Spyreas G. 2019. An assessment of the vulnerability of Illinois' rarest plant species to climate change. Castanea 84:115-27.